A TEXTBOOK
OF
ELECTRIC POWER
DISTRIBUTION AUTOMATION

A TEXTBOOK OF ELECTRIC POWER DISTRIBUTION AUTOMATION

By

DR. M.K. KHEDKAR
Dean (Administration) &
Professor (Electrical Engg.)
Visvesvaraya National Institute
of Technology, Nagpur
Maharashtra

DR. G.M. DHOLE
Professor (Electrical Engg.)
Shri Sant Gajanan Maharaj
College of Engineering, Shegaon
Maharashtra

UNIVERSITY SCIENCE PRESS
(An Imprint of Laxmi Publications Pvt. Ltd.)
An ISO 9001:2015 Company

BENGALURU • CHENNAI • GUWAHATI • HYDERABAD • JALANDHAR
KOCHI • KOLKATA • LUCKNOW • MUMBAI • RANCHI
NEW DELHI

A TEXTBOOK OF ELECTRICAL POWER DISTRIBUTION AUTOMATION

Printed and bound in India
Typeset at : Goswami Associates, Delhi.
First Edition : 2010, *Reprint* : 2014, *Edition* : 2017
ISBN : 978-93-80386-21-8

Branches

✆	Bengaluru	080-26 75 69 30
✆	Chennai	044-24 34 47 26
✆	Guwahati	0361-254 36 69
✆	Hyderabad	040-27 55 53 83
✆	Jalandhar	0181-222 12 72
✆	Kochi	0484-405 13 03
✆	Kolkata	033-40 04 77 79
✆	Lucknow	0522-430 36 13
✆	Ranchi	0651-224 24 64

PUBLISHED IN INDIA BY

UNIVERSITY SCIENCE PRESS
(An Imprint of Laxmi Publications Pvt. Ltd.)
An ISO 9001:2015 Company
113, GOLDEN HOUSE, GURUDWARA ROAD, DARYAGANJ,
NEW DELHI - 110002, INDIA
Telephone : 91-11-4353 2500, 4353 2501
www.laxmipublications.com
info@laxmipublications.com

C—12351/016/08
Printed at: Ajit Printers, Delhi.

Contents

Preface

Modern society is heavily dependent on uninterrupted and reliable availability of electricity at a high quality. There is an infinite array of services that just cannot function without a highly reliable source of electric power. Add to this, the mind-boggling number of domestic consumers whose life comes to a standstill if electric supply is disrupted. Thus, the importance of electricity distribution automation cannot be overemphasized.

Visvesvaraya National Institute of Technology, Nagpur introduced the course on "Electric Power Distribution" from last two years. This was in accordance with the recommendations from the BoS Committee, while revising the curriculum. The same has also been introduced to the Undergraduate course at final year, looking into the need for advanced technology in the field of Power Systems. This course is also offered at other Universities to mention few like JNTU, SK University, Burla Engineering College, etc.

However, no single book having the contents of the curriculum is available for teaching the course; it was thought of preparing the suitable learning material for the same. Accordingly, after having an exhaustive reference both from journals and websites relevantly mentioned at the end of each chapter, the complete structure of the learning material has been developed.

There is also a considerable variation in the syllabus content of this subject in different universities. An attempt has been made to include a variety of topics and prepare a comprehensive book on the subject. Finally, this book titled "A Textbook of Electric Power Distribution Automation" is the outcome of the experience of authors over the years.

We lay no claim to the original research in preparing the book. Liberal use of the materials available in works of eminent authors has been made. What we may claim, in all modesty, is that we have tried to fashion the vast amount of material available from both primary and secondary sources into a coherent body of description and analysis.

The organisation of the chapters is flexible and permits the instructor to mould them to particular needs of the curriculum and class. If necessary, some of the advanced topics may be bypassed without loss of continuity. The style of writing is amenable to self study.

Though the book's audience consists mainly of students, the practising engineers, interested in learning the fundamental concepts will also find it useful.

We are thankful to Mr. Mel George, our student, from the bottom of our heart, for the help he rendered in editing the contents of this book and also for his timely and useful suggestions. We are also thankful to Mr. Gokul Iyer, Mr. Karthik R. and others who helped us a lot.

The authors will gratefully receive suggestions and comments from the readers for improvement of the book.

—Authors

Chapter 1 INTRODUCTION

Power shortage in India is endemic. Despite of heavy investments in the power sector and additional power generation every year, there appears to be no chance of the shortage syndrome easing up in the near future. Barring one or two states with hydro potential, all other states and Union Territories have perpetual power deficits. There seems to be no respite from these crippling deficits as seen from the figures that are published from time to time.

In the power sector, it has almost become a ritual to insistently clamour for more generation. This school of thought which is predominantly in majority, believes in classical concept of load management which emphasizes continuous and copious supply of energy to all consumers at all times. This ideal proposition could be achieved, only if unlimited resources are available at our command. But, in actual practice, these resources are not only limited, but also scarce.

In India, large portion of the public resources are being spent on the power sector (about 30%). The installed generation capacity grew from 1300 MW to 95000 MW in the post independence period *i.e.*, about 73 fold increase. Still there is shortage of 18% during peak hours & 10% on an average. This means that India should invest more than Rs. 1, 50,000 crore to bridge the gap between demand and supply by enhancing the generation capacity installed.

Due to financial constraints, the electricity utilities are forced to look towards alternatives with more efficient use of electricity.

1.1 POWER SECTOR REFORMS IN INDIA

They were started from early 1992, but the true speed was felt only after establishment of regulators under the Regulatory Commission Act, 1998. With a view to improve the functioning of State Electricity Boards, the Government promulgated the State Electricity Regulatory Commission Act for establishment of CERC, at the national level and SERC in the States for rationalization of tariff.

The Electricity (Amendment) Act 1998 was passed with a view to make transmission as a separate activity for inviting greater participation in investment from public and private sectors.

1.1.1 Electricity Act, 2003

It is seen as a Bible of power sector reforms and the instrument to achieve the "power for all" goal set for 2012, having the potential to create a conducive business environment in the Electricity market. It has also thrown up new challenges and at the same time opened up newer opportunities to private players.

The aim is to:

- Consolidate the laws relating to Generation, Transmission and Distribution,
- Trading and use of electricity, generally for taking measures conducive to development of electricity industry, promoting competition therein,
- Protecting interests of consumers and supply of electricity to all areas,
- Rationalization of electricity tariff,
- Ensuring transparent policies regarding subsidies,
- Promotion of efficient and environmentally benign policies,
- Constitution of CEA, Regulatory Commission and Establishment of Appellate Tribunal etc.

TABLE 1.1 The Reform Status in Indian Power Sector

S. No.	Achieved so far	What it didn't
1	Transparency in Tariff Determinant	T&D losses did not improve significantly
2	Tariff Rationalization	Metering of all consumers not achieved
3	Reduction in cross-subsidy	Tariff increase for politically sensitive consumer segment has been difficult
4	Some improvement in operational efficiency	Promised subsidy by state government was delayed/deferred
5	Reassessment of T&D losses leading to greater transparency	Power shortage scenario continues
6		Slow private investment in power generation

(Source: Dr. Anoop Singh, "*Emerging Market Scenario in Indian Power Sector; An Assessment of Reforms*", Dept. of IME, IIT Kanpur, 2004)

TABLE 1.2

Region	Requirement (MU)			Peak Load (MW)		
	2006–07	2011–12	2016–17	2006–07	2011–12	2016–17
Northern	220820	308528	430000	35540	49674	69,178
Western	224927	299075	396000	35223	46825	61966
Southern	194102	262718	355000	31017	42061	56883
Eastern	69467	90396	117000	11990	15664	20416
N-Ern	9501	14061	21000	1875	2789	4134
Island	280	444	608	60	94	130
All India	719097	975222	1927000	115705	157107	212707

(Source: V.S. Verma, Member (Planning), CEA, New Delhi, Presentation on "*Power on Demand by 2012*").

1.1.2 Government's Vision of 'Power for All' by 2012

- Share of the private sector to be increased from 10% to 16.5%.
- Generating capacity to be increased to 212,000 MW.
- Hydel contribution to be increased to 30%.
- T & D losses to be reduced to 13%.
- 100% rural electrification.
- Industrial tariff brought down by half to support global competitiveness.

1.2 OBJECTIVES OF THE BOOK

Electrical engineering is an old subject but it is now rejuvenated with several new developments. This text treats the entire spectrum of distribution automation, from basic equipments to SCADA systems, load forecasting to network reconfiguration and remote metering to energy management. However, it is ironic that this also increases the possibility of disturbances by way of the malfunctioning of the automation system itself. This is possibly an area where automation tends to become an art. The engineer has to strike a balance between the threat perception and the security offered.

The book is focussed on teaching the fundamental concepts and the related design aspects of Distribution Automation (DA). Written in a clear, simple and down-to-earth style, the state of the art text offers electrical engineering students and practicing engineers a stimulating presentation that is both friendly and refreshingly simple. Most of the topics are introduced in interesting ways. The presentation is at a level which even students can grasp. Several sections of the book will also be helpful to practicing engineers.

Owing to the comprehensive nature of the book, there is more material than that which can be covered in a semester course. It is hoped that teachers and students would be able to make a judicious selection of the topics.

1.3 SYNOPSIS OF ALL CHAPTERS

Chapter 2 deals with load forecasting. Accurate load forecasting is necessary since these future load characteristics determine the location, size and timing of future facilities. They form the basis for development of strategic system plans. Advantages and disadvantages of short term and long term load forecasting have been highlighted.

Various objectives, functions and benefits of DA have been dealt in *Chapter 3*. DA requirements like the communication requirements, RTU's communication interface, and sectionalizing switches etc. has been discussed.

Chapter 4 describes a typical SCADA system. Components and functions of SCADA and moreover the need to integrate DA with SCADA has been considered. Chapter 4 also deals with the difficulties involved in implementing distribution automation and protocols used in it.

Deregulation of the utility industry is creating a situation where utilities must automate and acquire more information in order to remain competitive. Hence *Chapter 5* deals with need for substation automation. Functions and benefits of SA have also been mentioned. Recent protection practices and trends of SA have been discussed.

Chapter 6 introduces the optimal switching device placement in radial distribution network. Types of sectionalizing switches and their benefits have been highlighted. Bellman's optimality principle, which can be used for determining the optimal number and position of switching devices for radial network, has also been introduced. Network reconfiguration is necessary to minimize the outage area, reduce system losses and balance loads to avoid overloading of network elements. Minimum nodal voltage method for feeder reconfiguration has also been described in *Chapter 6*.

The distribution lines frequently experience faults. It then becomes necessary to locate and isolate the faulty section. The various types of faults and methods used to locate these have been described in *Chapter 7*. Process of restoration has also been discussed.

With rising energy costs there is an increased need to identify utility usage and wastage. Hence *Chapter 8* deals with benefits of automatic meter reading (AMR) system. The components of AMR have also been described here.

In the last decades of the 20th century, India has been facing an acute crisis due to the generation being lesser than the demand. Various types of energy management and energy audit as a tool to save energy have been discussed in *Chapters 9 & 10*.

Various costs are associated with owning, operating and maintaining the power distribution system. A suitable pricing and charging structure is required to achieve economy in power distribution. *Chapter 11* deals with the various costs and terms like return on investment and depreciation. Methods used for economic appraisal have also been discussed.

Electrical distribution network is complex in nature and extendable to any limit. So, all conventional methods almost fail to give exact global solution to the problem. Hence *Chapter 12* employs AI based approach in order to get exact feasible solution to the existing problems in the distribution network.

Chapter 13 deals with the Urban and Rural Distribution system. Primary and secondary distribution system, high voltage system and single wire earth return system have been described in this chapter.

Chapter 14 deals with optimal switching device placement in radial distribution system. Sectionalizing switches, needs, types and additional circuitry have been discussed in this chapter.

Chapter 15 deals with Communication Technology. Different techniques used for communication, and difficulties in implementing communication technology have been discussed in this chapter.

1.4 GENERATION, TRANSMISSION AND DISTRIBUTION

Electricity, produced and delivered to customers through generation transmission and distribution system, constitutes one of the largest consumer markets in the world. Numbers vary for individual utilities, but the cost of electricity is approximately 50% fuel, 20% generation, 5% transmission and 25% distribution.

Reliable electric power systems serve customer loads without interruptions in supply voltage. Generation facilities must produce enough power to meet customer demand. Transmission system must transport bulk power over long distances without overheating or jeopardizing system stability. Distribution systems must deliver electricity to each customer's service entrance. In the context of reliability, generation, transmission and distribution are referred to as functional zones.

Each functional zone is made up of several subsystems. Generation consists of generation plants and generation substations. Transmission consists of transmission lines, transmission switching stations and transmission substations and sub transmission systems. Distribution systems consist of distribution substations, primary distribution systems, distribution transformers and secondary distribution systems. A simplified representation of an overall power system and its subsystems is shown in Fig. 1.1.

1.4.1 Generation Subsystems

Generation Plants produce electrical energy from another form of energy such as fossil fuels, nuclear fuels or hydropower. Generation Substations connect generation plants to transmission lines through a step-up transformer that increases voltage to transmission levels.

1.4.2 Transmission Subsystems

Transmission Systems transport electricity over long distances from generation substations to transmission or distribution substations. Transmission Switching Stations serve as nodes in the transmission system that allow transmission line connections to be reconfigured. Transmission Substations are transmission-switching stations with transformers that step down voltage to sub transmission levels.

Sub transmission Systems transport electricity from transmission substations to distribution substations.

1.4.3 Distribution Subsystems

Distribution Substations are nodes for terminating and reconfiguring sub-transmission lines plus transformers that step down voltage to primary distribution levels. Primary Distribution Systems deliver electricity from distribution substations to distribution transformers. Distribution Transformers convert primary distribution voltages to utilization voltages. Secondary Distribution Systems deliver electricity from distribution transformers to customer service entrances.

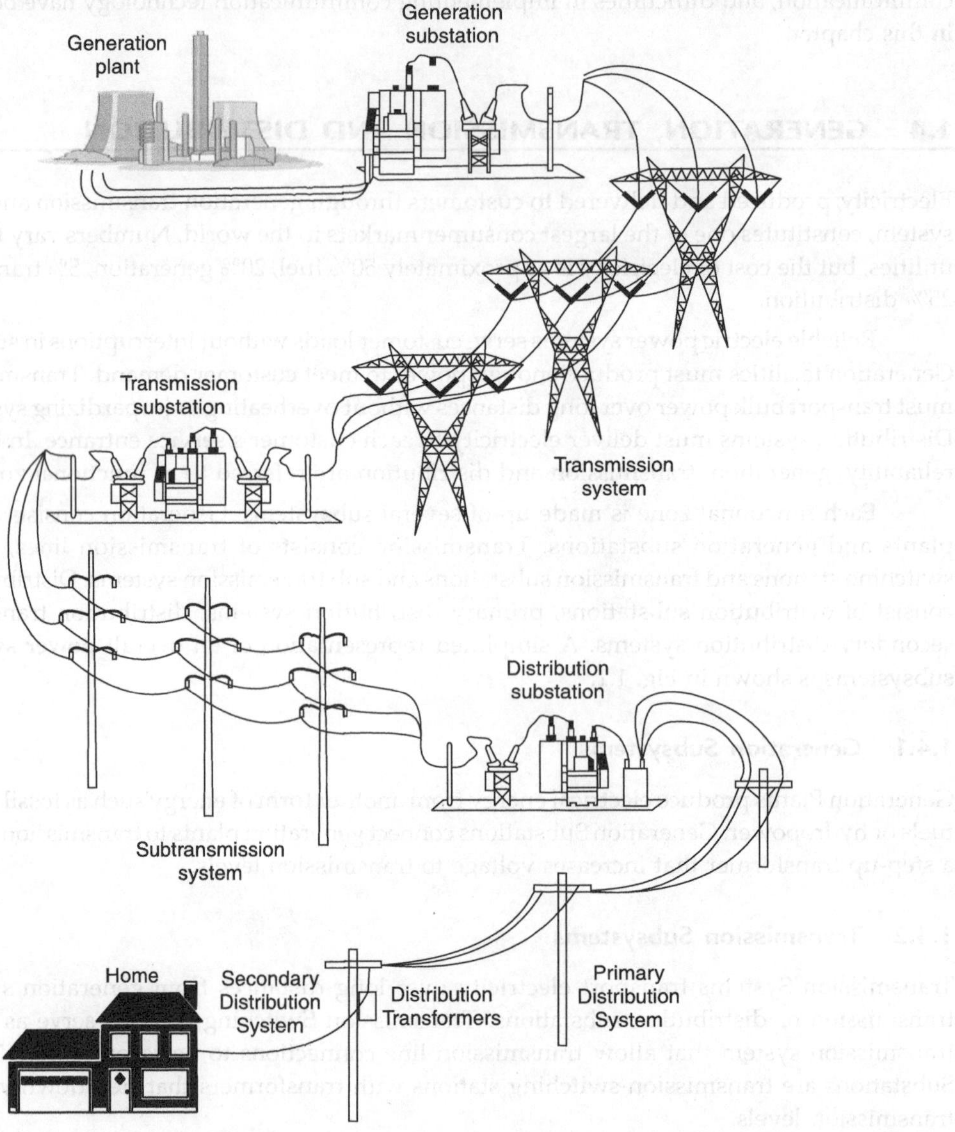

FIGURE 1.1 **Electric Power Subsystems**

(*a*) Generation of ac (hydro, thermal, nuclear, diesel, non-conventional sources etc.) 0.4/0.44, 6.6, 10.5, 11, 13.8, 15.75, 21 and 33 kV.

(*b*) (*i*) Tieline (interconnecting line between two power systems/areas) and transmission: 220, 400, 500, 750, 765, 800 kV, etc.

(*ii*) High voltage dc transmission: 500, 800, 1400 kV, etc. for stable power flows over long distances beyond 800–1600 km.

(*c*) High voltage subtransmission: 33, 66, 110, 132 kV, etc.

(*d*) High voltage primary distribution: 3.3, 6.6, 11, 22, 33, 66, 132 or 220 kV. 11 and 22 kV is the main primary distribution voltage used in India. The higher voltages are used for large consumers.

(*e*) Low-voltage (secondary) distribution, ac, 415/240 V and 433/250 V, three-phase four-wire; 240/250 V single phase to neutral; 440/220 V (from HT SWER).

In practice, a power system is complicated. It has a number of power stations of different types, interconnected by a system of tielines, transmission lines, subtransmission lines and distribution networks to supply different types of loads to various consumers.

1.5 T&D SYSTEM'S MISSION

A T&D system's primary mission is to deliver power to electrical consumers at their place of consumption and in ready-to-use form. The system must deliver power to the customers, which means it must be dispersed throughout the utility service territory in rough proportion to customer locations and demand. This is the primary requirement for a T&D system, and one so basic it is often overlooked—the system must cover ground—reaching every customer with an electrical path of sufficient strength to satisfy that customer's demand for electric power.

That electrical path must be reliable, too, so that it provides an uninterrupted flow of stable power to the utility's customers. Reliable power delivery means delivering all of the power demanded, not just some of the power needed, and doing so all of the time. Anything less than near perfection in meeting this goal is considered unacceptable - 99.9 % reliability of service may sound impressive, but it means nearly nine hours of electric service interruption each year, an amount that would be unacceptable in nearly any first-world country.

Beyond the need to deliver power to the customer, the utility's T&D system must also deliver it in ready-to-use form—at the utilization voltage required for electrical appliances and equipment, and free of large voltage fluctuations, high levels of harmonics or transient electrical disturbances.

In many countries, utilization standards vary from 230 to slightly over 250 volts, at either 50 or 60 cycles AC. But regardless of the utilization voltage, a utility must maintain the voltage provided to each customer within a narrow range centered within the voltages that electric equipment is designed to tolerate.

A ten percent range of delivery voltage throughout a utility's service area may be acceptable, but a ten percent range of fluctuation in the voltage supplied to any one customer is not. An instantaneous shift of even three percent in voltage causes a perceptible and to some people, disturbing, flicker in electric lighting. More important, voltage fluctuations can cause erratic and undesirable behavior of some electrical equipment.

Thus, whether high or low within the allowed range, the delivery voltage of any one customer must be maintained at about the same level all the time—normally within a range of three to six percent—and any fluctuation must occur slowly, such stable voltage can be difficult to obtain, because the voltage at the customer end of a T&D system varies inversely with electric demand, falling as the demand increases, rising as it decreases. If this range of load fluctuation is too great, or if it happens too often, the customers may consider it as poor service.

Thus, a T&D system's mission is to:

1. Cover the service territory, reaching all customers.
2. Have sufficient capacity to meet the peak demands of its customers.
3. Provide highly reliable delivery to its customers.
4. Provide stable voltage quality to its customers.

And of course, above everything else, to achieve these four goals at the very lowest cost possible.

1.5.1 The "Laws of T&D"

The complex interaction of a T&D system is governed by a number of physical laws relating to the natural phenomena that have been harnessed to produce and move electric power. These interactions have created a number of "truths" that dominate the design of T&D systems:

1. It is more economical to move power at high voltage. The higher the voltage, the lower the cost per kilowatt, to move power upto any distance.
2. The higher the voltage, the greater the capacity and the greater the cost of otherwise similar equipment. Thus, high voltage lines, while potentially economical, cost a great deal more than low voltage lines, but have a much greater capacity. They are only economical in practice if they can be used to move a lot of power in one block—they are the giant economy size, but while always giant, they are only economical if one truly needs the giant size.
3. Utilization voltage is useless for the transmission of power. The 120/240 volt single-phase utilization voltage used in the United States or even the 250 volt/416 volt. Three-phase used in "European systems" is not equal to the task of economically moving power more than a few hundred yards. The application of these lower voltages for anything more than very local distribution at the neighborhood level results in unacceptably high electrical losses. Severe voltage drops, and astronomical equipment cost.
4. Power is more economical to produce in very large amounts. Claims by the advocates of modern distributed generators notwithstanding there are significant economy of scale in generation—large generators produce power more economically than small ones. Thus, it is most efficient to produce power at a few locations utilizing large generators.
5. Power must be delivered in relatively small quantities at low (120 to 250 volt.) voltage level. The average customer has a total demand equal to only 1/10,000th or 1/100,000th of the output of a large generator.

1.6 BASIC DISTRIBUTION SYSTEMS

The six basic distribution systems used by utilities (shown in Fig. 1.2) are as follows:

(*i*) **Radial.** A radial system is connected to only one source of supply (see Fig. 1.2(*a*)). It is exposed to many interruption possibilities—the most important of which are those due to overhead line or underground cable failure, or transformer failure. Each event may be accomplished by a long interruption. It has lower reliability. Both components (feeder and transformer) have finite failure rates and such interruptions are expected and statistically predictable. Feeder breaker reclosing or temporary faults are likely to affect sensitive loads. This system is suitable for small loads.

(ii) **Primary Loop.** A great improvement over a radial system is obtained by arranging a primary loop, which provides power from two feeders (see Fig. 1.2(*b*)). This is also called open ring system. Power flow to the consumer is by way of a single path at any one time from either side of the loop, depending upon the open/close status of sectionalizers and reclosers. The loop is normally operated with the sectionalizer switch open. Any section of the feeder can be isolated without interruption, and primary faults are reduced in duration to the time required to locate a fault and do the necessary switching to restore service. Each line of the loop must have sufficient capacity to carry all the load. The additional line exposure tends to increase the frequency of faults, but not necessarily the faults per consumer. Sensitive loads are affected by reclosing under temporary fault conditions.

(iii) **Primary Selective.** It uses the same basic components as in the primary loop. Each transformer can have supply from two sources (see Fig. 1.2(*c*)). High voltage automatic switching is provided ahead of the consumer's transformer. In the event of loss of feeder, transfer to the second feeder is automatic and the interruption duration can be limited to two or three seconds. Each service now represents a potential two feeder outage if the open switch fails, but under normal contingencies, service restoration is rapid and there is no need to locate the fault, as with the loop switching system, reliability is high. It also offers little advantage to sensitive load-like computer problems caused by temporary faults. This scheme is normally used for large, essential or continuous process industrial consumers.

(iv) **Secondary Selective.** This system uses the two transformers, each from a separate primary feeder and with low voltage switching (see Fig. 1.2(*d*)). The load is generally divided between two LT buses, with both transformers continuously energized. The tie switch on the secondary tie bus is normally open and is interlocked with the secondary feeder's switches. This system is commonly used for industrial plants and institutions like hospitals. Primary operational switching is eliminated. Duplicate Transformers virtually eliminate the possibility of a long interruption due to failure. Each transformer and feeder must have sufficient capacity to supply the entire load. Transfer is automatic upon loss of voltage in either feeder with static switching equipment. Sensitive equipment can be effectively served. Reliability is better than in the primary selective system because of the additional redundancy of transformers.

(v) **Spot Network.** Maximum service reliability and operating flexibility for most loads are obtained by use of the network, using two or more transformer units in parallel. It is similar to the closed ring system. The low voltage bus (spot network bus) is continuously energized by all units operating in parallel. Automatic disconnection of any unit is obtained by sensitive reverse (directional) power relays in the protection unit (see Fig. 1.2(*e*)). If one feeder develops a fault, it is isolated by the protection on that feeder.

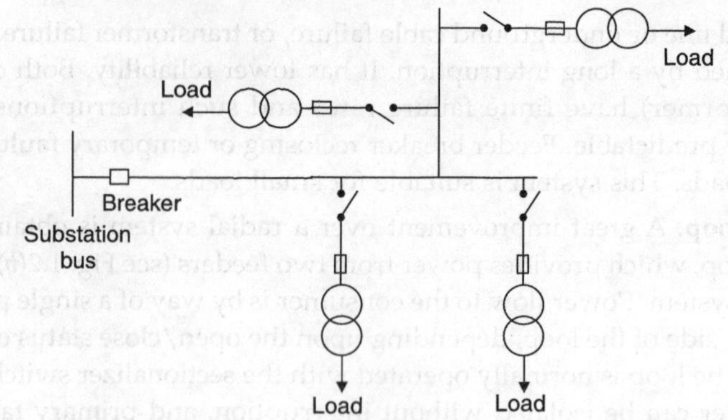

(*a*) Radial

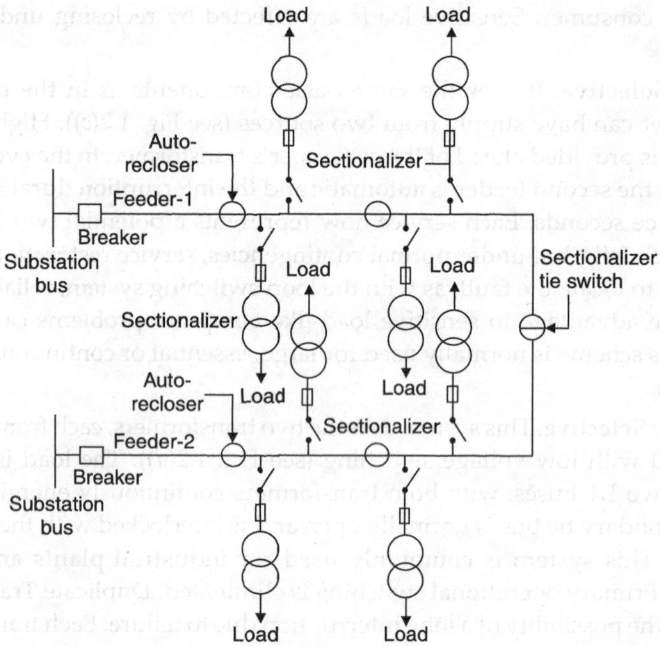

(*b*) Primary loop

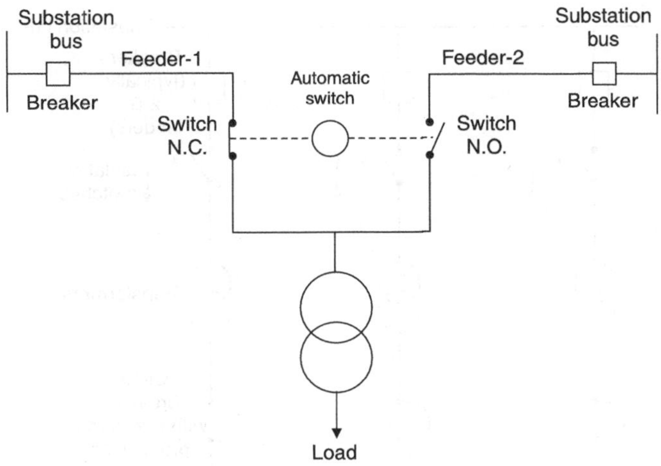

(*c*) Primary selective

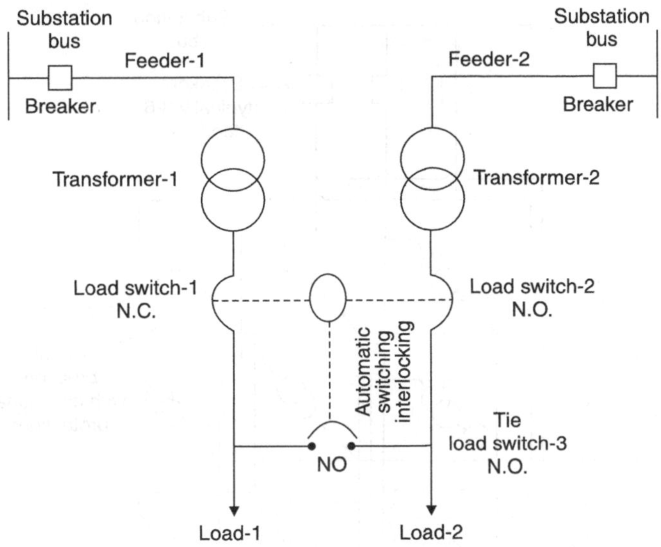

(*d*) Secondary selective

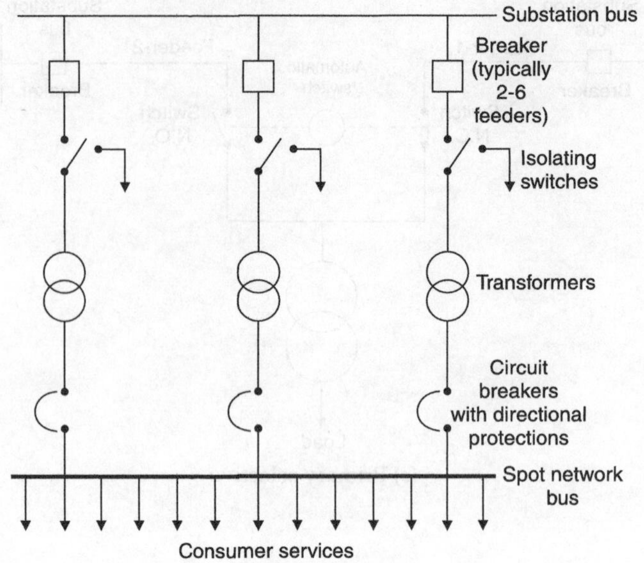

(*e*) Spot network

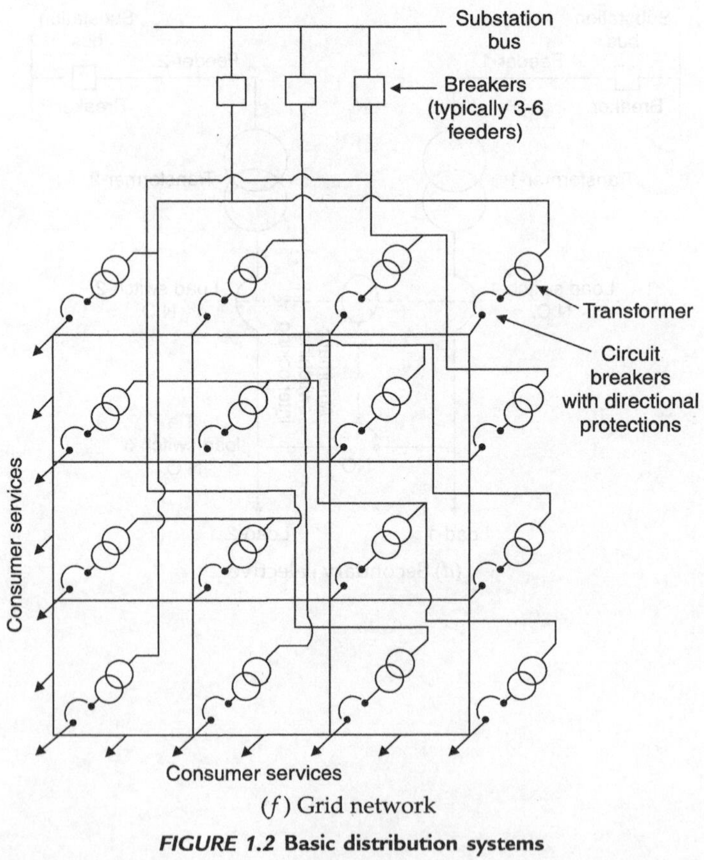

(*f*) Grid network

FIGURE 1.2 **Basic distribution systems**

Switching for the maintenance of the primary feeder can be done without consumer interruption. Spot networks are generally used in metropolitan or high load density areas for large continuous process industries and essential services load such as water works etc. A spot network, because of its multiplicity of feeders and transformers, is highly reliable. Momentary and long duration outages are almost non-existent.

(*vi*) **Grid Network.** Grid networks provide maximum reliability and operating flexibility. These networks are the most economical and effective method in serving the high density loads in metropolitan cities. The grid is simultaneously supplied from several feeders. In the grid network, no consumer outage is caused by switching off the primary feeder for scheduled maintenance. Voltage regulation is improved since power flow to the consumer is through several transformers operating in parallel (see Fig. 1.2(*f*)). The grid can handle abrupt load changes and disturbances associated with large motor starting, without severe voltage dips or surges. A strong grid network is sufficiently stiff and a fault in one unit does not disrupt voltage outside the sensitive load tolerance limits.

1.7 SUMMARY

The power sector reforms were started in India in 1992. The Central Government evolved the APDRP (Accelerated Power Development and Reforms Programme) to reduce T&D losses, achieve reliable power supply and to bring about reforms in the distribution sector. Recently, the Electricity Act 2003 was also enacted. Despite all these measures, we are still far away from what we set out to achieve *i.e.*, to be a developed country. The main cause of worry is the distribution sector and very high T&D losses.

There are basically 3 subsystems in the power system viz. generation, transmission and distribution. These are also referred to as functional zones. The distribution systems commonly used by power supply utilities are radial, primary loop, primary selective, secondary selective, spot network and grid network, each of which have their pros and cons. There also exists a wide difference between urban and rural distribution.

The basic mission of a T&D system is to deliver electric power to consumers at their place of consumption and in a ready-to-use form. However, this mission is subject to certain rules called as the "Laws of T&D".

LONG TYPE QUESTIONS

1. What are the aims of the Electricity Act, 2003?
2. What is the Indian Government's vision of 'Power for all'?
3. What is the mission of a T & D System?
4. What are the Laws of T & D?
5. Write a short note on the following types of distribution systems:

   (*a*) Radial

   (*b*) Primary Loop

(*c*) Primary Selective
(*d*) Secondary Selective
(*e*) Spot Network
(*f*) Grid Network.

MULTIPLE CHOICE QUESTIONS

1. The true speed of the power sector reforms in India was felt only after the establishment of regulators under
   (*a*) Regulatory Commission Act, 1988 (*b*) Regulatory Commission Act, 1996
   (*c*) Electricity Amendment Act, 1998 (*d*) Electricity Amendment Act, 1996.
2. The Bible of Power Sector reforms is
   (*a*) Electricity Act, 2000 (*b*) Electricity Act, 2003
   (*c*) Electricity Amendment Act, 1998 (*d*) Regulatory Commission Act, 1998.
3. Which of the following voltages are suitable for long distance DC transmission?
   (*a*) 500 kV (*b*) 650 kV
   (*c*) 850 kV (*d*) 1430 kV.
4. The primary mission of a T & D System is
   (*a*) to generate power (*b*) to reduce losses
   (*c*) to deliver power to electrical consumers at their place of consumption
   (*d*) none of the above.
5. Which of the following distribution systems is connected to only one source?
   (*a*) Primary Selective (*b*) Radial
   (*c*) Secondary Selective (*d*) Spot Network.
6. An electrical distribution system has to be implemented such that a fault at one part does not affect other parts of the system. Which of the following networks would be suitable for this?
   (*a*) Primary Selective (*b*) Radial
   (*c*) Grid Network (*d*) Spot Network.

REFERENCES

1. *"Electric Power Distribution"* by A. S. Pabla. 5th Edition.
2. *"EHVAC & HVDC Transmission and Distribution"* by S. Rao.
3. www.ntpc.co.in
4. www.indiapoweronline.com.

Chapter 2 DISTRIBUTION SYSTEM PLANNING

2.1 INTRODUCTION

Power System load forecasting is an important subject both in power system planning as well as in on-line power system operation. In recent years, significant variations between supply of and demand for electricity have led to violent price swings on the wholesale market. To minimize our vulnerability to market volatility, we want to optimize our load forecasting accuracy so that we can enhance distribution system efficiency by helping ensure that we use what we have scheduled and schedule what we will use. More accurate forecasting capability can also enable us to negotiate more profitable power purchases between utilities, retail companies, power generators and in Independent System Operator (ISO/RTO) scheduling environments. Inaccurate schedules can result in energy imbalance charges.

Load forecasts form the basis for development of strategic system plans. Accurate load forecasting is critical since the future load characteristics determine the location, size, and timing of future facilities. ABB is unique in the power industry in its capability to perform load forecasts using small-area land-use based simulation techniques. A key aspect of spatial load forecasting is the tie between land use/development and electrical load. This tie is achieved by assigning each small area of development to the substation serving the area. This tie allows the forecast to be calibrated to actual load levels as well as "rollup" of load growth to components of the electrical system.

- These load forecasts, which can be forecasts of either peak MW requirements during a period (typically 1 hour) or total MWH requirements for that period, are used by system dispatchers and operations analysts to control and to plan power.
- The annual peak load forecasts can be used by system planners for development of a system expansion plan to meet long terms needs by leveraging current system expansions to address longer term capacity requirements.

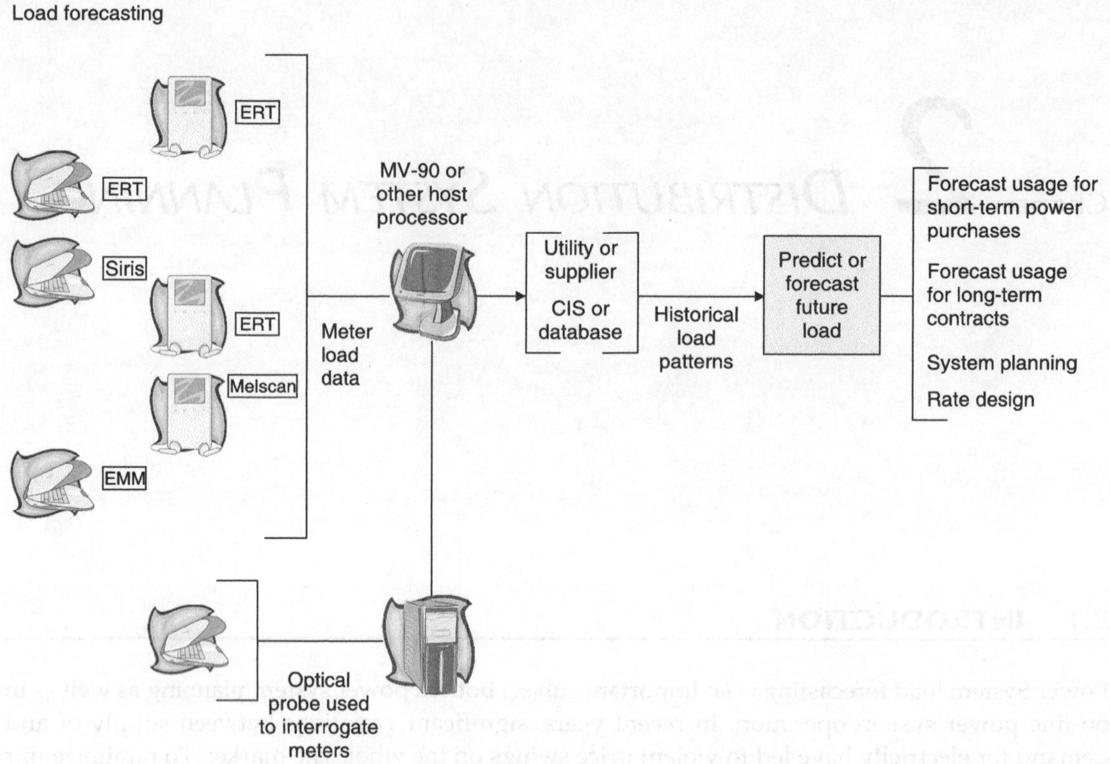

FIGURE 2.1 **Load Forecasting block diagram**

- An accurate System Load Forecasting function, used to calculate short-term electric load forecasts, is essential component of any Energy Management System (EMS) system operations.
- Forecasts of hourly loads for up to one week ahead are necessary for scheduling functions such as Hydro-Thermal Coordination and Transaction Evaluation and for short-term analysis functions such as Dispatcher Power Flow and Optimal Power Flow.
- The Hydro-Thermal Coordination function requires hourly system load forecasts for the next day or the next week in order to establish the hourly schedules for generation resources that will minimize the system operating cost, subject to reliability requirements, operational constraints and other limitations imposed on system operation.
- The Transaction Evaluation function requires the hourly system load profile for its study period in order to determine the costs and effects of proposed capacity and energy exchanges.

2.2 SHORT-TERM LOAD FORECASTING

Short-Term Forecast has to be simple, speedy and robust. Short-term load forecasting for operational planning is necessary not only for arranging the operational plan of the next 24 hours which generator of which power plant has to be put into operation, or how much load it should share, or which generator has to be taken out of commission but also for meeting the load change of the coming hours. It takes time for the boiler of the thermal power plant to admit more fuel and increase combustion to produce more steam for increasing power generation. The load dispatcher has to have a fairly good estimate of the increase of load so as to decide how to share the increment of load adequately and economically between the operating generation units, or power plants.

Generally the daily load change follows the daily pattern, which is different between week days and weekend days. But it subjects to weather change rather sharply, and the weather change is more often than not to be predicted not very accurately either in magnitude or in time.

The load-forecasting algorithm can go to great complexity, but still it cannot be better than the experienced engineer's judgments. It is felt that what the operator needs for on-line forecasting is fairly accurate software, simple to handle. For on-line load forecasting, speed is, of course, of paramount importance.

Short-term load forecasting plays an important role in electric power system operation and planning. An accurate load forecasting not only reduces the generation cost in a power system, but also provides a good principle of effective operation. ARIMA model and transfer function model are applied to the short-term load forecasting by considering weather-load relationship. For four types of customers in Taiwan power (Taipower) system, residential load, commercial load, office load and industrial load customers, the summer ARIMA model transfer function model has been derived to precede the short-term load forecasting during one week.

- ARIMA model and transfer function model for customer short-term load forecasting during one week in distribution system.
- The ARIMA model completely considers the dynamic process of data series, time delay variables and the auto correlation of residuals in order to achieve precise load forecasting.
- Furthermore, the transfer function model introduces the interpreted variable of temperature into univariate ARIMA model to enhance the load forecasting ability.
- The load customers are sampled from load survey system and classified into four types of commercial load, office load, residential load, and industrial load customers.
- The hourly load patterns of various customers are derived by statistic procedure, which processes power consumption data over one year. Besides, the conventional regression model is applied to be reference for comparison.
- The proposed models for these four types of load customers are applied to forecast future short-term load trend to demonstrate the effectiveness of proposed approaches.
- ARIMA and transfer function models can provide satisfactory load forecasting accuracy. Finally, the transfer function model considered the effect of temperature and can achieve better accuracy of load forecast than ARIMA model and can provide guidance of short-term load forecasting to Taipower Company.

The SLF algorithm produces forecasts for system loads for the current day and the following day using three models: a peak load forecast model, an hourly load forecast model and a load update model. The peak and hourly models contain separate submodels for the summer period (May through October) and the winter period (November through April).

2.3 LONG-TERM ENERGY FORECASTING

In power system planning, no matter whether the planning period is 5 years or 50 years, the planner has to forecast the load for all these years within the planning period. The accuracy of prediction influences greatly the capital investment and is therefore very important. When numerous energy sources and their development schemes are involved there could be a great number of construction alternatives to fit the estimated load growth and there could be different orders, or Sequences of putting them into commission.

2.3.1 Disadvantages of Long-Term Forecasting

- Long-term Load Model is Complicated and Strongly Influenced by Weather.
- Long-term load forecasting for a planning period of, say 25–50 or more years has such a far remote target that the objective is usually of very uncertain nature.
- Yet certain possibilistic estimate is required. It has to be re-estimated and revised as often as necessary, semi-annually or even quarterly or whenever some new important issues are foreseen.
- In such uncertain cases, planner needs to take new data or make new assumptions to forecasting.
- Then different planners will make different assumption in the revision process, and when the decision-makers Involved are numerous, the number of cases to be studied might be enormous. It is very desirable to do the Forecast as quickly and as easy as the planner wishes.
- Hence, speed of computation and easiness of use of the software by the planner ate more important than the unwarranted accuracy under the condition of great number of cases to be handled and the wide range of variations.

The arrangement of different generation expansion plans to meet the different load growth pattern, which is not a simple straight line can suggest a number of hydro-thermal combinations, which will vary with load pattern. Thus, the load pattern is also important. In comparing the different suggested generation expansion schemes, we have to compare not only the capital cost but also the productioin cost in operation, so not only the annual load curve has to be estimated, but also the seasonal variation has to be taken into account, in order that the generation expansion scheme can fit the load growth in an optimal way.

For the load forecast—to be accurate enough for production costing is not a simple matter. It is known to be greatly influenced by weather conditions. Temperature is a most important factor, but humidity, cloud land wind are also to be considered. Hence load forecast has to depend on weather forecast, whose accuracy is difficult in time, space and magnitude, not to speak of storm or unpredictable abnormal weather changes. Forecasting annual peak load demand and annual energy demand for a number of years ahead has a vital role in the context of generation, transmission and distribution network planning in a power system.

- An approach to forecasting energy is that based on an end-use model. In the end-use method, energy forecast is based on the estimated energy use of all load categories in different sectors.

- Typical sectors include residential, industrial, commercial and service sectors. Within each sector, load categories or components are identified. For example, in residential sector, the load categories include air-conditioning and heating, lighting, refrigeration, washing, and cooking.
- Average energy use in each load category is then needed in the forecast. The growth rates in the use of individual load components are also required.

In principle, the end-use approach is an appealing one. However, there is difficulty in collecting data which is essential for use in the forecast. Extensive customer surveys are required in categorising load components and their use.

2.4 DISTRIBUTION OF POWER

Electricity is modern society's most convenient and useful form of energy. Without it the present social infrastructure would not at all be feasible. The increasing per capita consumption of electricity through out the world reflects the growing standard of living of people .The optimum utilization by the society of this form of energy can be ensured by an effective distribution system.

The distribution system is part of the system between transmission and the consumer service point.

It contains:

- Sub transmission circuit in voltage ratings, usually between 33 kV and 220 kV, which delivers energy to distribution substation.
- The distribution substation, which converts the energy to lower primary system voltage for local distribution.

2.5 DISTRIBUTED ENERGY SUPPLY SYSTEM

There are three types of consumption units–interruptible (*e.g.*, blower, shredder), controllable (*e.g.*, heat- and cold-storage devices) and not interruptible loads. In addition to electrical energy, heat or steam is produced in co-generation units. The thermal energy is tapped and passed to the consumers for production processes or heating.

This is conditional on efficient communication between system control center and generation units, storage units and consumption units, as well as on an innovative distributed energy management system with forecasting (for weather, generation and load), operations planning, online-optimization as well as generation control and load control.

2.6 TECHNOLOGICAL FORECASTING

- Technology forecasting is the process of discovering and communicating probable technical capabilities in order to make better decisions, gain competitive advantage and prevent surprises.

- Technology forecasting concentrates on discovering the direction, rate, and nature of the changes taking place in technologies.
- The tools available for accomplishing these tasks fall into four general categories of techniques: surveillance, trend analysis, expert opinion, and integrative.
- The techniques within the four are aimed at discerning the probable capabilities at the three levels of technology (direct, supportive, and enabling).
- The types of techniques used and the efforts employed to forecast are dependent on a series of both internal and external factors.

2.6.1 Technological Forecasting Methods

- Numerical Data-Based Technological Forecasting Techniques
  - Trend Extrapolation
  - Trend Extrapolation, Qualitative Approaches
  - Growth Curves
  - Envelop Curves
  - Substitution Model
- Judgment-Based Technological Forecasting Techniques
  - Monitoring
  - Network Analysis
  - Morphological Analysis
  - Relevance Trees
  - Delphi Method
  - Cross-Impact Analysis.

LONG TYPE QUESTIONS

1. Explain the following terms:
   (*a*) Load Forecasting
   (*b*) Short-term Load Forecasting
   (*c*) Long-term Load Forecasting.
2. Explain the ARIMA Model and Transfer Function Model of Load Forecasting.
3. Explain Technological Forecasting.
4. Describe the methods of Technological Forecasting.
5. State the disadvantages of long-term forecasting.
6. Explain the End-use approach of energy forecasting.

MULTIPLE CHOICE QUESTIONS

1. Short-term load forecasting is necessary for arranging the operational plan of
   (*a*) 24 hours (*b*) 24 days
   (*c*) 24 months (*d*) 24 years.
2. ARIMA and Transfer Function Models are applied to the Short-term Load Forecasting by considering
   (*a*) Customer demand (*b*) load demand
   (*c*) Weather conditions (*d*) relationship between weather and load.
3. When the planning period is 5–50 years, it is called
   (*a*) Short-term load forecasting (*b*) Long-term load forecasting
   (*c*) Technological forecasting (*d*) None of the above.
4. Long-term forecasting is influenced by
   (*a*) period of planning (*b*) customer demand
   (*c*) weather (*d*) load demand.
5. Technological forecasting helps in
   (*a*) discovering and communicating technical capabilities
   (*b*) make better decisions
   (*c*) gain competitive edge (*d*) all of the above.
6. Which of the following approaches can be adopted for Energy Forecasting?
   (*a*) ARIMA Model (*b*) Transfer Function Model
   (*c*) SLF Algorithm (*d*) End Use Model.
7. The distribution system consists of
   (*a*) sub-transmission circuit (*b*) distribution sub-station
   (*c*) both (*a*) and (*b*) (*d*) none of the above.
8. Heat and cold storage devices are examples of
   (*a*) interruptible consumption unit (*b*) non-interruptible consumption unit
   (*c*) controllable interruptible consumption unit
   (*d*) none of the above.
9. Which of the following is a major disadvantage of the End-use approach of energy forecasting?
   (*a*) extensive customer surveys are required (*b*) it is dependent on humidity
   (*c*) it is dependent on temperature (*d*) none of the above.

Chapter 3 DISTRIBUTION AUTOMATION

3.1 PROBLEMS WITH EXISTING DISTRIBUTION SYSTEM

The problems of Indian Distribution network and loads are described here in relevance to DAS.

- The annual load growth is high (around 10% to 12 %) necessitating setting up of additional system facilities every year to meet the growing demand.
- The power losses in distribution network are high: around 15% of total energy generated and reduction of loss is profitable.
- The distribution equipment is fully loaded or overloaded causing high equipment failure rate.
- The power factor of system is poor, requiring efficient reactive power management.
- The agricultural pump sets contribute for 80% of peak demand in rural area and 40 % of total energy sales and therefore load control of agricultural pump sets is very attractive.
- Voltage profile of system is unsatisfactory.
- The number of breakdowns and momentary interruptions on distribution feeder are high.

3.2 NEED FOR DISTRIBUTION AUTOMATION

The recent regulatory changes have caused emergence of new issues for the utilities in managing the power system and concern in retaining the customers. As distribution utilities shift from non-profit public utilities to profit-driven business enterprises, the question how to maintain and improve service reliability while keeping electricity rates lower and protect shareholders' interests becomes more difficult to answer. The trend that service reliability is going downwards is not uncommon as seen in some of the utilities. Reliability has a cost and regulators have to provide incentives to improve the system reliability. Besides the changing of the utility business, the emergence of new sensitive loads such as e-business data centre, Internet Service Provider (ISP) and industry with sensitive manufacturing process demanding uninterrupted and stringent quality of power supply. These new loads not only add on the system peak demand about 5–13% and grow fast, but also require high standard of service reliability and power quality.

The evolutionary growth in microprocessor based devices and telecommunication equipment and network have brought the possibility of integrating protection, control, metering, automation and monitoring system cost effectively. This will considerably improve system reliability, quality of supply, and customer service and defer capital investment on system expansion. Distribution automation and system monitoring meet the demands and requirements in improving service reliability at a lower cost.

The real need for advanced distribution automation is associated with the growing demand for the reliability of power supply and desire for optimized network conditions in normal and emergency operations. One of the most important aspects of distribution automation system is featured by the electric utility dispatchers' ability to remotely monitor, coordinate and operate distribution components.

3.3 CHARACTERISTICS OF DISTRIBUTION SYSTEM

The Distribution system is part of the system between transmission and the consumer service point. It contains:

- Sub transmission circuits in voltage ratings, usually between 33kV and 220 kV, which delivers energy to distribution substations.
- The distribution substation converts the energy to lower primary system voltage for local distribution and usually improve facilities for voltage regulation of primary voltage.
- Primary circuits of feeders usually operating in the range of 11kV to 33kV supplying the load in a well defined geographical area.
- Distribution transformers usually installed on poles or on pads or near the consumers sites, which transform the primary voltage to the utilization voltage, usually 230/400V.
- Secondary circuits at utilization voltage, which carry energy from the distribution transformers along the streets etc.
- Service lines which deliver the energy from secondary circuits to the consumer premises.

3.4 DISTRIBUTION AUTOMATION

Distribution Automation (DA) can be defined as automatically monitoring, protecting and controlling switching operations through intelligent electronic devices to restore power service during fault by sequential events and maintain better operating conditions back to normal operations. If we add Demand Side Management (DSM) functions, such as Automatic Meter Reading (AMR) and Load Management (LM), into distribution automation, it comes a broad definition of Distribution Automation.

Distribution automation can be divided into three categories, such as transformer substation, feeder, and customer premises automations. Usually the distribution automation on substation and feeder are integrated to share common monitoring and controlling equipment and devices. The DA functions on transformer substation include substation control such as bus voltage control, circulating current control, bus load balancing and overload control, and transformer protection such as bus fault isolation, and transformer isolation. The DA functions on feeder mainly include feeder automatic switching and feeder voltage/VAR control. The former deals with fault identification, fault location, fault isolation and service restoration. The latter contains capacitor placement and voltage regulator control. The DA functions on the customer premise could be in a broad range covering load control, real-time price signalling, remote meter reading and billing, etc.

3.4.1 Objectives of Distribution Automation

The following are objectives for the Distribution Automation System (DAS):

(*i*) Reduce peak load and power losses to overcome prevailing power shortages and defer construction of distribution facilities.

(*ii*) Improve the reliability of supply by reducing the number and duration of outages, and improve the quality of service.

(*iii*) Improve the financial performance of the utility by improved cash flow, safeguarding revenues, and preventing theft of power.

3.4.2 Distribution Automation Functions

The selection of functions depends on network conditions, load characteristics, tariff policies, and other socio-economic factors. The following are few Distribution Automation functions:

Load Control

- Shedding of loads for which time of operation is not important like pumping loads to reduce system peaks and prevent overloading of distribution equipment.

Substation Automation

- Monitoring the status of substation equipment
- Monitoring voltages and circuit loading
- Supervisory control of substations
- Collection of substation's historical data.

Feeder Automation

- Fault isolation and service restoration
- Remote circuit switching
- Collection of feeder historical data.

Feeder Volt and VAR Control

- Remote control of LTCs, regulators, and capacitors.

Automatic Meter Reading (AMR) at large customers

- Time of use metering
- Detection of meter tampering and theft
- Monitoring of meter for proper operation
- Remote service disconnects and reconnects.

3.4.3 Benefits of Distribution Automation

The benefits of distribution automation are indicated below:

Substation Automation

- Avoid equipment damage by timely recognition of problems
- Prevent outages and thereby improve continuity of service by remotely monitoring and recording data on the state of the system
- Prevent outages and thereby reduce revenue losses by remotely monitoring and recording data on the system
- Defer construction of new distribution facilities as a result of accurate monitoring of circuit loads. This allows the rating of the substation transformers to be exceeded for a short time
- Save labour costs by unattended operation of substations
- Facilitate substation engineering and maintenance by historical data collection.

Feeder Automation

- Reduce outage time and improve the continuity of service.
- Reduce outage time and reduce revenue losses.
- Defer construction of new facilities.
- Reduce losses.
- Release generation and transmission capacity through loss reduction.
- Reduce labour costs by reducing staff responsible for routine and emergency switching operations.
- Facilitate network engineering and maintenance by historical data collection.

Feeder Volt and VAR Control

- Defer construction of new distribution facilities.
- Reduce energy losses by real-time control of capacitor switches.

- Release generation and transmission capacity through reduced losses.
- Improve voltage profile by VAR control and power factor correction.

Automatic Meter Reading (AMR)

- Improve cash flow by faster billing.
- Reduce labour costs by reassigning staff responsible for meter reading.
- Reduce theft through tampering detection, remote connect and disconnect of service, and comparison of metered delivery with circuit power measurements.

Load Control

- Avoid damage to distribution transformers and other equipment by introducing diversity in the pumping loads and, if necessary, shedding of all pump sets.
- Defer construction of new distribution facilities.
- Reduce losses.
- Release generation and transmission capacity.
- Eliminate need to shed complete feeders as an emergency measure for load relief. This will result in better service and reduced loss of revenue.
- Reduce outages caused by equipment overload.

3.4.4 Definitions/Important Terms in Distribution Automation

- **Man machine interface (MMI).** It is the interface between man and technology for control of the technical process. The computer system at master control centre or central control room integrates with RTU over the communication link with its transmission protocol, acquires the remote substation or distribution transformer/ feed a data and transfers the same to the computer system for MMI. Fig. 3.1 shows the flow diagram for

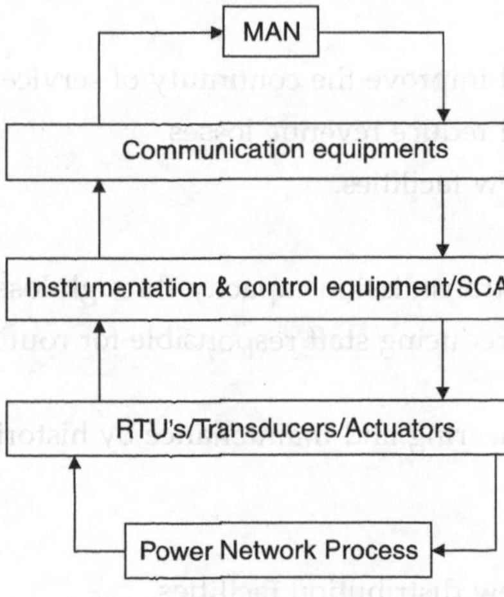

FIGURE 3.1 **Flow diagram for man machine power interface**

man machine interface for the power system. The software is confined in a manner that makes the single line diagram of the distribution system. The entire system can be monitored and controlled from the screen.

- **Automation Switching Control.** Some of these are as follows:
  1. *Photoelectric Outdoor Lighting Controls:* Local electromagnetic relay or electronic and/or thermal/time delay relay with photoelectric controls.
  2. *Line or Capacitor Switching Controls*: Local control by time, temperature, current, voltage and Var., remote control by VHF/UHF radio.
  3. *Line Post Sensors*: SCADA and local feeder monitoring for load switching and fault information.
  4. *Faulted Circuit Indications*: Fault location through use of on site LED, fluorescent flag, and remote SCADA indication for single-phase or 3-phase overhead, underground or pad mounted transformers.
  5. *Radio Switches*: Peak power demand reduction through VHF/UHF radio switching of consumer or feeder load.
- **Information Technology.** It includes administrative computing and all 'end-user' computing of a business and technical nature. The term extends to smart or intelligent, programmable electronic devices used in power operations, from the generation of electricity (with computer based distributed plant control systems) to power distribution automation (including applications of such technologies as automated mapping systems, SCADA, EMS etc.)
- **Management Information Systems.** Most often defined in the power utility industry, as the department responsible for administrative computing systems and operations.

3.5 FEEDER AUTOMATION

Feeder automation includes:

- Feeder automatic sectionalizing for fault location, fault isolation, service restoration, and feeder reconfiguration
- Feeder remote point voltage control
- Feeder reactive power control
- Substation reactive power control
- Substation transformer load balancing
- Cold-load pickup
- Automatic reclosing

Reconfiguring of feeders helps to minimize overloads and balance loads on substation transformers, allows the utility to defer capital expenditures for adding substation and feeder capacity. In addition, reconfiguring feeders can reduce feeder losses. This has a two-fold benefit.

First, less fuel must be burnt to generate the energy for losses. Distribution transformer losses often are about 5% of the system energy output and that a reduction of about 10% can be achieved by reconfiguring feeders. Second, the generation and transmission capacity required to produce and transmit the loss energy can be used for more productive purposes. Although it is possible that generation and transmission capacity additions could be deferred, and the potential savings are large, it is unlikely that any deferrals would be realized, due to the already present delays in obtaining permits and completing construction.

When automatic feeder sectionalizing is used for fault location and isolation, O & M costs for crews to find and isolate the fault will be reduced. Service restoration will be faster, enabling the utility to resume earning revenue for energy sales and reducing the cost of the outage to the customers. The number of customer complaints will also be reduced, along with the cost of responding to them. It is generally understood, however, that the cost of an outage to a residential customer is small, particularly for outages of short duration.

Feeder remote point voltage control, feeder reactive power control, and substation reactive power control can all result in reduced losses and a reduced number of damage claims based on low voltage.

A cold load pickup function can reduce O & M costs to the extent that crews would not be required to sectionalize feeders in order to reenergize them after long outages.

The major cost element of feeder automation is not the functions themselves, but the field devices and communication systems. Because the object of the feeder automation functions is to control feeder devices, the control capability of feeder SCADA must be installed

The typical distribution automation system composed of master station and the distribution equipment for feeder and consumer automation system is as shown in Fig. 3.2. The details of equipment are given in Table 3.1.

TABLE 3.1 Typical equipment of distribution automation system

Master Station	Distributed Equipment	
	Feeder Automation	Consumer Automation
Work station	RTU	RTU for transformer
Communication Equipment	Remote controlled SF_6 gas switch	RTU for consumer
Line Sensor	Communication media	Communication media
Console or MMI		

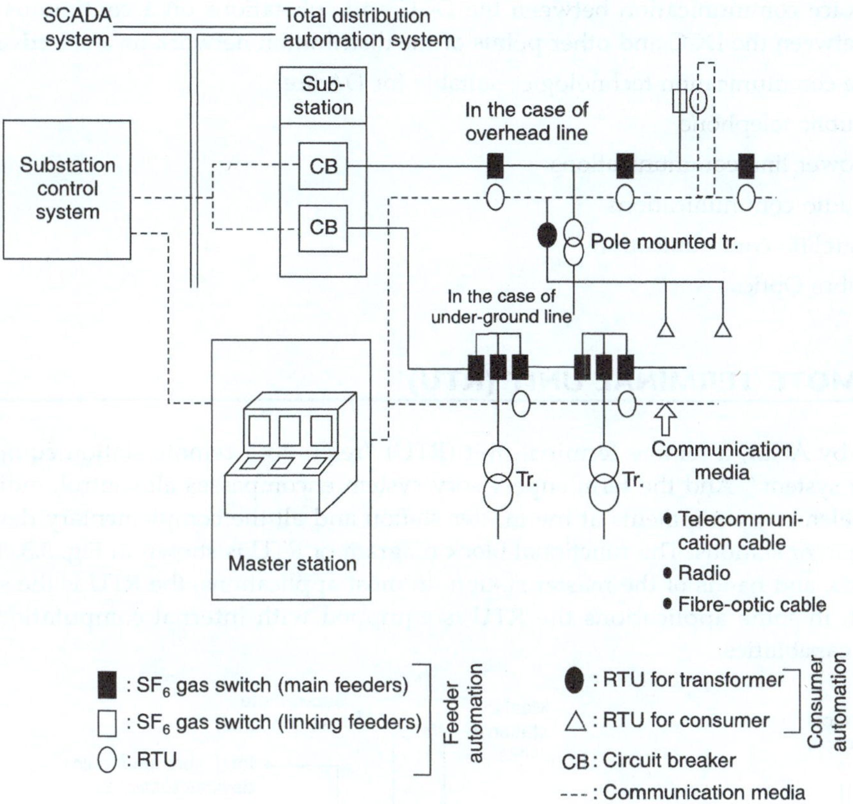

FIGURE 3.2 **Typical distribution automation system**

3.6 COMMUNICATION REQUIREMENTS FOR DA

Communication is very important to Distribution Automation because DA relies on the coordinated effort of various elements in the system. The following DAS communication requirements have to be met:

- Two-way data communications between a Distribution Control Centre (DCC) and substations for SCADA.
- Two-way data communications between the DCC and distribution transformers for load control of pump set groups, and for monitoring of voltages and loading.
- Two-way data communications between the DCC and switches located on the primary distribution network for capacitor control and network reconfiguration.
- Two-way data communications between the DCC and large customers with automatic meter reading (AMR).
- One-way data communications between the DCC and agricultural pump sets for load control.

- Voice communication between the DCC and substations on a continuous basis, and between the DCC and other points of the distribution network on a selective basis.

Three communication technologies suitable for DA are:

(*i*) Public telephone
(*ii*) Power line communications
(*iii*) Radio communications
(*iv*) Satellite communications
(*v*) Fibre Optics.

3.7 REMOTE TERMINAL UNIT (RTU)

As defined by ANSI, a remote terminal unit (RTU) "refers to a remote station equipment of a supervisory system." And the term supervisory system encompasses all control, indicating and associated telemetry equipments at the master station and all the complementary devices at the remote station or stations. The functional block diagram of RTU is shown in Fig. 3.3. The RTU is the eyes, ears, and hands of the master station. In most applications, the RTU is the slave of the master; but, in some applications the RTU is equipped with internal computational and/or optimizing capabilities.

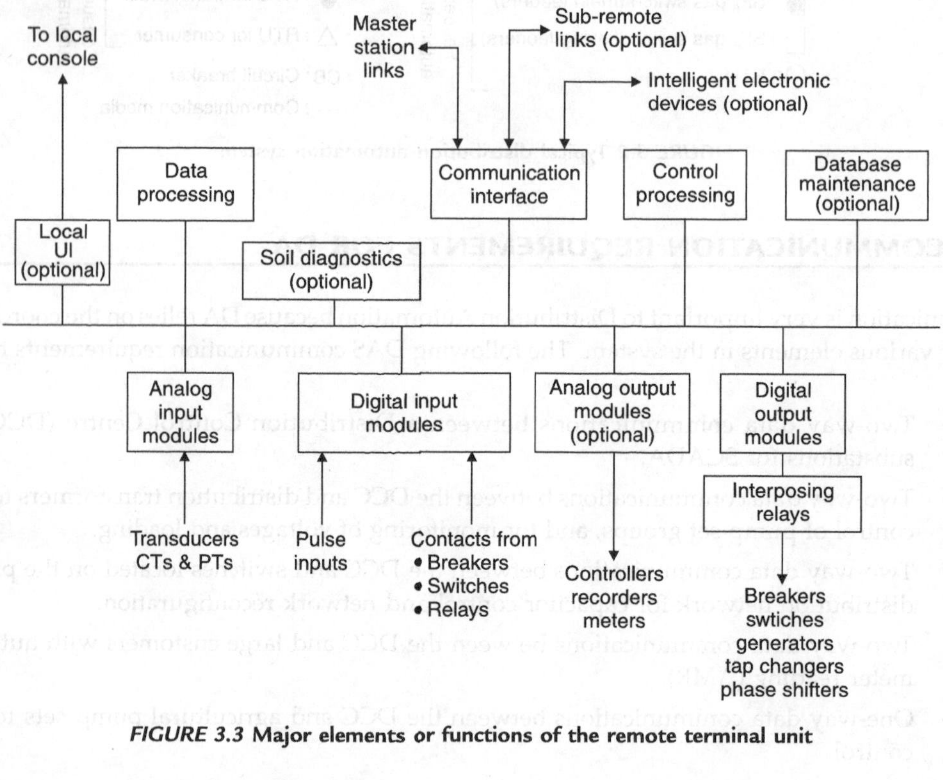

FIGURE 3.3 **Major elements or functions of the remote terminal unit**

3.7.1 Communication Interface

Power system RTUs interface to a large number of communication systems. In fact, many utilities use several communication media on a single SCADA system. To both the RTU and the master, most of these communication media are technically equivalent to four-wire telephone circuits. Modern RTUs interface to most of these communication systems through built-in modems, with selectable communication speed options ranging from 300 to 1,200 bits per second.

Master Station Links: The RTU communication subsystem is responsible for interpreting messages from the master and formatting messages to be sent to the master. Thus, this subsystem handles all protocol and communication security-related functions.

Intelligent Electronic Devices (IEDs). Microprocessors have proven their capability, reliability, and cost effectiveness in RTUs; and they are providing similar characteristics for other substation control, protection, and instrumentation devices. As microprocessors are added to these other devices, they are programmed to monitor and/or calculate the various system parameters required to determine their operational performance. When this data is supplied via digital transfer to the RTU, a great many hardware devices are eliminated from the substation.

3.7.2 Data Processing Master Stations

The data processing subsystem consists of collecting and reporting two types of data: digital and analog. Digital quantities are obtained by monitoring external contacts, *e.g.*, circuit breakers, switches, and pulse generators. Analog quantities are normally obtained from transducers which provide either current or voltage output in proportion to the primary parameter being measured. Some RTUs are now collecting 0 – 10 VAC analog quantities from post insulators, and other devices, in proportion to primary voltages and currents.

3.7.3 Digital Input Modules

Most SCADA systems monitor and collect four types of digital data or contact position information:

- Current status or state.
- Current status with memory detect, which is the number of contact changes since the last report to the master.
- Sequence-of events (SOE) log (tabulation of contact changes with time of occurrence).
- Accumulator value, which is a count of the number of contact closures over a period of time. The most common use of accumulators is in association with energy meter pulse generators, where the pulse stream is proportional to energy flow.

SOE logging requires the RTU to time-tag events to within 1.0 millisecond, requiring both an accurate clock and an interrupt driven processor. Therefore, the RTU must be designed and equipped to support time synchronization and time keeping. Some systems provide synchronization from the master; other systems install radio or satellite time receivers in each RTU. SOE I/o points must be stored in two files, one for immediate change of state feedback to the master and the other a time-tagged log. The SOE log are used by system planning and system protection engineering to verify that the protection system work as designed.

Accumulator input values are usually stored as 16- or 32-bit words, depending upon the accuracy required. These values are handled in a variety of ways by different utilities. Some utilities operate accumulators like a car's odometer; *i.e.*, the accumulator keeps counting until it returns to zero. Other utilities read and reset the accumulator at the end of each time period. Some utilities freeze all accumulators at the same point in time; so time consistent values are available over the entire system. The freeze command tells the RTU to store the current accumulator count in a buffer for reporting on the next poll, and then continue counting. It is important that no counts be lost during the freeze process.

3.7.4 Digital Output Modules

Control output points usually support separate TRIP and CLOSE relays, a select-before-operate (SBO) security provision, and circuitry to insure that one, and only one, control relay in the RTU is operated. The one-and-only-one safeguards may include any combination of logical and analog circuits, sensing the relay coils if possible, and a coil supply which is current-limited for only one relay coil. The SBO feature provides for an end-to-end system check, whereby the master sends a point selection message and receives a check back message, before sending an operate message and receiving a verification message. The selection point number is stored and the check back point number should be regenerated from as close to the relay coil as may be activated by the selection logic. The activate transaction should include a repeat of the point number, which the RTU should compare against its selected point before operating the relay. A timer should allow an un operated point to remain selected for only a few seconds. All control logic should be reset upon completion of an operation or upon detection of any error. Direct-operate output points can be provided for issuing incremental RAISE/LOWER commands to power apparatus controllers. These output points support single or multipoint selection in response to a single command from the master. The output is in the form of either a single contact closure of variable duration, or a variable number of closures of fixed duration.

3.7.5 Analog Input Modules

As mentioned previously, analog input values are normally obtained from transducers as milliamp or voltage values. In order to process the values in the RTU and SCADA system, they must be converted to digital values using an analog-to-digital (A/D) converter.

3.7.6 Analog Output Modules

Analog output is the least often used of the basic RTU I/O functions. The most common type is that which produces a constant milliamp signal proportional to a digital quantity specified in a command from the master. Volt or millivolt outputs are sometimes required, but the current standards are 0 to plus or minus 1 mA into 0 to 10K Ohms, and 4 to 20 mA into 0 to 600 Ohms.

3.7.7 Control Processing

Implementation of closed-loop control is achieved by the RTU comparing a measured quantity to a set point downloaded from the master. If the measured quantity is too low, a control action is taken to increase the value. If the measured value is too high, a control action is taken to reduce the value. Usually, a dead band or permissible error is defined to limit excessive control action. Closed-loop control has been used on load-tap changing transformers, voltage regulators, hydroelectric gates, etc.

3.7.8 Termination Subsystem

The termination subsystem provides the interface between the RTU logic subsystem and the electric power subsystem. The functional requirement for the termination subsystem is to protect the RTU from all substation environmental conditions. A high voltage substation can provide a very noisy and otherwise hazardous electrical environment for what might seem to be rather sensitive electronic circuitry. But with foreknowledge of the potential problems appropriate counter measures, overall good design practice and thorough RTU design evaluation on the part of the manufacturer the user can expect reliable service.

3.7.9 Communication Technology

3.7.9.1 Communication Infrastructure and Technologies

Communication is very important for a SCADA system. It is required to do its functions based on the data sent by the utility. For this purpose data is to flow from the utility to SCADA and from SCADA to utility. This requires a reliable and qualitative communication system. A general communication system for SCADA is given in Fig. 3.4.

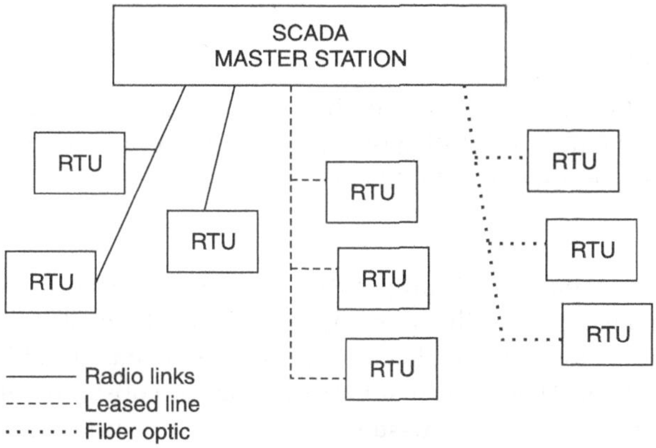

FIGURE 3.4 **Typical SCADA system communication architecture**

Communications technology developments continue to advance and drive the SCADA industry into new directions. Because of these communication developments, the substation SCADA system will assume a larger role in the utility enterprise. The advent of fibre optics, direct satellite broadcast, and third-party digital service providers has made high data rate transmission economical.

There are many communication methods available. Evaluation of different communication systems for data communication between Distribution control centre (DCC) and any point on the distribution network is required at the planning stage. The fundamental requirements for communication infrastructure are:

(*i*) Determination of system average message rate;

(*ii*) Capacity to handle the requisite amount of data and multitasking;

(*iii*) Data throughput and system response times should meet various application requirements;

(*iv*) It should allow for network growth and added applications.

The communication may be used individually or combined.

3.7.9.2 Public Telephone Communication (PTC)

Dial up telephone is suitable for remote metering as the requirement is infrequent data transfer and less costly. Leased or dedicated telephone is suitable for substation automation as frequent scanning is required. But cost is very high. Public telephone is not suitable for feeder automation.

3.7.9.3 Radio Communication

Radio communications are similar to PLCC systems in the fact that these are also generally owned by the utility. But, they have the important advantage of independent operation, which is not based on the status of the Power System. The general used media for the radio communication are UHF (Ultra High Frequency) and VHF (Very High Frequency). Radio communication suffers from the drawback of limited channels available for the communication purpose. Generally used Radio technologies are:

UHF Point-to-Point Radio

This is the communication technology which is widely used for two way point to point communication between the utilities. With a well established infrastructure this technology can be of good use for distances ranging from 30–40Km.

UHF MARS

UHF MARS is Ultra High Frequency Multiple Address Radio System. This was developed as a substitute for the dedicated phone lines and is widely used for SCADA/EMS/Distribution Automation applications. In this scheme there is a master radio station which communicates with several remote radio stations. This is most suitable media for substation and feeder automation. Multi address radio systems consist of master radios, which communicate with remote radios. This is two-way communication with baud rate of 9.6 KB. Each system uses two frequencies, one for master to remote & one for remote to master.

VHF Radio

VHF RADIO is Very High Frequency Radio. This is one way communication suitable for control of load switches mounted on pole tops. This is generally used for sending load control commands. Inexpensive receivers/switches are available for giving load control switching commands.

Cellular Radio

This is the system which is being widely used for AMR, DSM and DA. The main benefits of this system are wide-area broadcast and mentoring capability, and the independence from communication utility. This technology is considered to be the idealistic one for the field operations because of its high speed real-time and bidirectional data gathering systems control.

3.7.9.4 Fibre Optics

This is the technology which has become the most widely accepted choice in the modern era. This has been possible due to the explicit downward cost trend in terms of product as well as installation and maintenance cost, as it offers both technical and commercial advantages over the conventional technologies that use metallic cables and radio links. The communication system mainly comprises of a trans receiver set which is connected through copper wires to the switching centre, where it is converted into a digital signal for easy handling. The main feature of this system is that it is not encountered by the electric interference. This in turn gives the advantage of high data rate and immunity from noise.

3.7.9.5 Satellite Communication

The main component of this type of communication technology is the VSAT (Very Small Aperture Terminal). The whole communication is dependent on the VSAT, VSAT Hub and the related data. The main feature which forms the strength for this technology is its reliability factor that is as high as 99.5%. Another main advantage of the technology is that it is very insensitive to distance. In this configuration the whole control is done by the earth station which is called the hub. The hub maintains a link between all the VSATs which are present in distant locations.

This technology is useful for load management controls. Satellite based VSAT (Very Small Aperture Terminal) network communication is used for power systems. It has a 99.5% reliability factor. A communication signal sent from a VSAT to the satellite at a particular frequency. The signal is amplified, shifted to a different frequency and sent back to the VSAT at the destination. A VSAT network is distance insensitive. One has to pay only for the cost of VSATs and service charges, which are very reasonable. The whole network can be monitored and controlled centrally. If there is any problem in the network it can be diagnosed from the hub. Primarily, there are three types of technologies for VSAT networks. The Government of India has approved the 'star' technology. In this configuration, at the centre of the network is a big earth station known as the hub with a dish diameter 6 – 11 meters. The control of the network is at the hub and it maintains a communication link between VSATs scattered all over the country or region. The configuration is therefore called the Star.

The other technology is the 'MESH' technology. The configuration permits a group of VSATs to communicate directly with other VSATs going through the central hub. In this type of configuration, any one of the VSATs functions as a hub and controls the whole network. This configuration enables better voice and video communication than the Star configuration. This technology has also been introduced in India. In the third type of configuration, there is point to point connection between the VSATs. In such a network, data, voice and video signals are carried directly between the locations.

In satellite communication, an extended C-band is used for communicating the data.

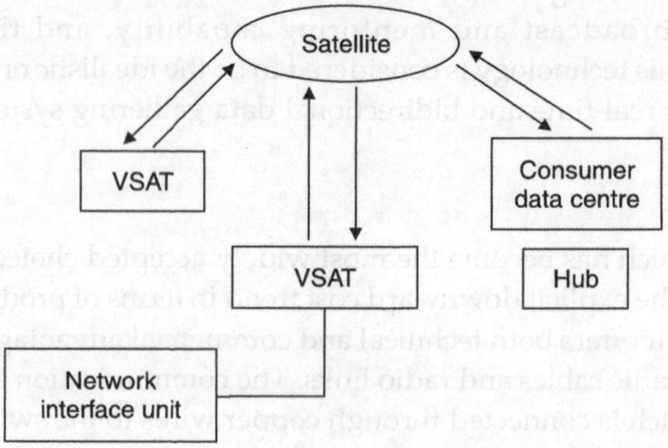

FIGURE 3.5 **Schematic diagram for VSAT Network**

LONG TYPE QUESTIONS

1. State and explain the various problems plaguing existing distribution systems.
2. Explain the need for Distribution Automation.
3. Define the following:
   (*a*) Distribution Automation
   (*b*) Demand Side Functions
   (*c*) Man-machine Interface
   (*d*) Remote Terminal Unit
   (*e*) Fibre Optics
   (*f*) VSAT
   (*g*) Power line communication
   (*h*) Radio communication
   (*i*) Satellite communication.
4. Enlist the objectives of Distribution Automation.
5. Enumerate the functions of distribution automation.
6. Explain the benefits of distribution automation as applied to:
   (*a*) Substation Automation
   (*b*) Feeder Automation
   (*c*) Feeder Volt and VAR Control
   (*d*) Automatic Meter Reading (AMR)
   (*e*) Load Control.
7. Why is man-machine interface important in distribution automation?
8. Explain how the above is achieved with the help of a neat flow diagram.
9. Explain how Information Technology is applied to distribution automation.

10. 'Reconfiguring of feeders has a two-fold benefit'. Explain.
11. Explain why communication equipments are important in distribution automation.
12. What are the major elements of an RTU?
13. Explain the following as applied to RTU:
    (*a*) Communication interface
    (*b*) Data processing master stations
    (*c*) Digital input and output variables
    (*d*) Analog input and output variables
    (*e*) Control processing.
14. Why is a good communication system essential for a distribution automation?
15. What are the fundamental requirements of a communication system?
16. What are the different radio communication techniques?
17. What are the advantages of fibre optics over conventional technologies?

MULTIPLE CHOICE QUESTIONS

1. The annual load growth in India is around
   (*a*) 5–10% (*b*) 10–12%
   (*c*) 50–60% (*d*) 20–30%.
2. What is the contribution of agricultural equipments to the peak demand in India?
   (*a*) 70% (*b*) 75%
   (*c*) 80% (*d*) 85%.
3. The use of automation and microprocessor based devices helps in
   (*a*) improved system reliability (*b*) quality of supply
   (*c*) customer satisfaction (*d*) all of the above.
4. Which of the following is the utilization voltage?
   (*a*) 11 kV (*b*) 440 kV
   (*c*) 440 V (*d*) 230 V.
5. Which of the following does the term 'feeder automation' include?
   (*a*) feeder remote point voltage control (*b*) sub-station reactive power control
   (*c*) automatic reclosing (*d*) all of the above.
6. Which of the following helps minimize overloads?
   (*a*) power factor control (*b*) reactive power control
   (*c*) reconfiguration (*d*) automatic reclosing.
7. The major cost involved in feeder automation is:
   (*a*) the distributed automation functions (*b*) the field devices
   (*c*) communication systems (*d*) both (*b*) and (*c*).
8. In a process the analog values in an RTU, which of the following can be used?
   (*a*) latch (*b*) buffer
   (*c*) A/D converter (*d*) D/A converter.

9. Radio communication is similar to which of the following systems?
   (*a*) Public telephone systems (*b*) PLCC systems
   (*c*) Satellite communication (*d*) None of the above.
10. Which of the following communication technologies can be used for two way communication between the utilities?
    (*a*) UHF MARS (*b*) VHF Radio
    (*c*) Cellular Radio (*d*) UHF point to point radio.
11. The major advantage of fibre optics communications is
    (*a*) it is not affected by electrical interference (*b*) high rate of data transmission
    (*c*) immunity from noise (*d*) all the above.
12. Which of the following VSAT networks is used in Indian satellite communication?
    (*a*) Mesh technology (*b*) Star technology
    (*c*) Both (*a*) and (*b*) (*d*) None of the above.

REFERENCES

1. M.V. Krishna Rao , Reuven Peri Kay , N. Clinard , Dr. C. Radha Krishna and J. V. Pandurangam "*Development and Evaluation of a Distribution Automation System for an Indian Utility*" IEEE Transactions on Power Delivery, Vol. 10, No. 1, January 1995 , pp 452–458.
2. William R Cassel "*Distribution Management Systems: Functions and Payback*" IEEE Transactions on Power Systems, Vol. 8, No. 3, August 1993, pp 796–801.
3. Roy Billinton, and Satish Jonnavithula "*Optimal Switching Device Placement in Radial Distribution Systems*" IEEE Transactions on Power Delivery, Vol. 11, No. 3, July 1996, pp 1646–1651.
4. David G. Hart, David Uy, James Northcote Green, Carl LaPlace and Damir Novosel, "*Automated solutions for Distribution Feeders*", IEEE Computer Applications in Power, Oct 2000, pp 25–30.
5. H. Lee Smith and Wayne R. Block, "*RTUs Slave for Supervisory Systems*", IEEE Computer Applications in Power, Jan 1993, pp 27–32.
6. Rong-Liang Chen and Shafi Sabir "*The Benefits of Implementing Distribution Automation and System Monitoring in the Open Electricity Market*" Canadian Conference on Electrical and Computer Engineering 2001, Toronto, pp 825–831.
7. A. S. Pabla, "*Electric Power Distribution*", Tata McGraw Hill, Fifth Edition, 2003.
8. H. Lee Smith and Thomas J. Modzelewski , "*Enhancing Energy Management Systems with Advanced RTU capabilities*" IEEE Computer Applications in Power, October 1989, pp 26–29.
9. Douglass L. Campbell, "*How Customer Need Focused the Development of a New Remote Terminal Unite Line*", IEEE Computer Applications in Power, July 1988, pp 9–11.
10. M Gans Conlog, A Division of AMIC Industries Limited, South Africa, "*Development of a Pole-Mounted RTU For USE on Rural Power Lines*", Power System Control and Management, 16–18 April 1996, Conference Publication No. 421, IEE, April 1996, pp 103–107.

Chapter 4 SCADA System

4.1 INTRODUCTION

SCADA is an acronym for Supervisory Control and Data Acquisition. It can be defined in many ways. Some of them are:

- SCADA is a computer system for gathering and analyzing real time data, which is used to monitor and control a plant or equipment.
- SCADA (supervisory control and data acquisition) is an industrial measurement and control system consisting of a central host or master (usually called a master station, master terminal unit or MTU), one or more field data gathering and control units or remotes (usually called remote stations, remote terminal units, or RTUs) and a collection of standard and/or custom software used to monitor and control remotely located field data elements.

These systems are generally used in industrial and engineering applications to monitor and control distributed systems from a master location. SCADA systems can be relatively simple, such as one that monitors environmental conditions of a small office building, or incredibly complex, such as a system that monitors all the activity in a nuclear power plant or the activity of a municipal water system. SCADA systems were first used in the 1960s.

SCADA is a very broad umbrella that describes solutions across a large variety of industries, including but not limited to the following:

- Electric power generation, transmission and distribution
- Oil and Gas Industry metering and control systems
- Environmental control systems
- Traffic signals
- Water management systems
- Mass transit systems
- Manufacturing systems.

4.2 BLOCK DIAGRAM

The block diagram of a general SCADA system is shown in Fig. 4.1. It shows all the connections in the system.

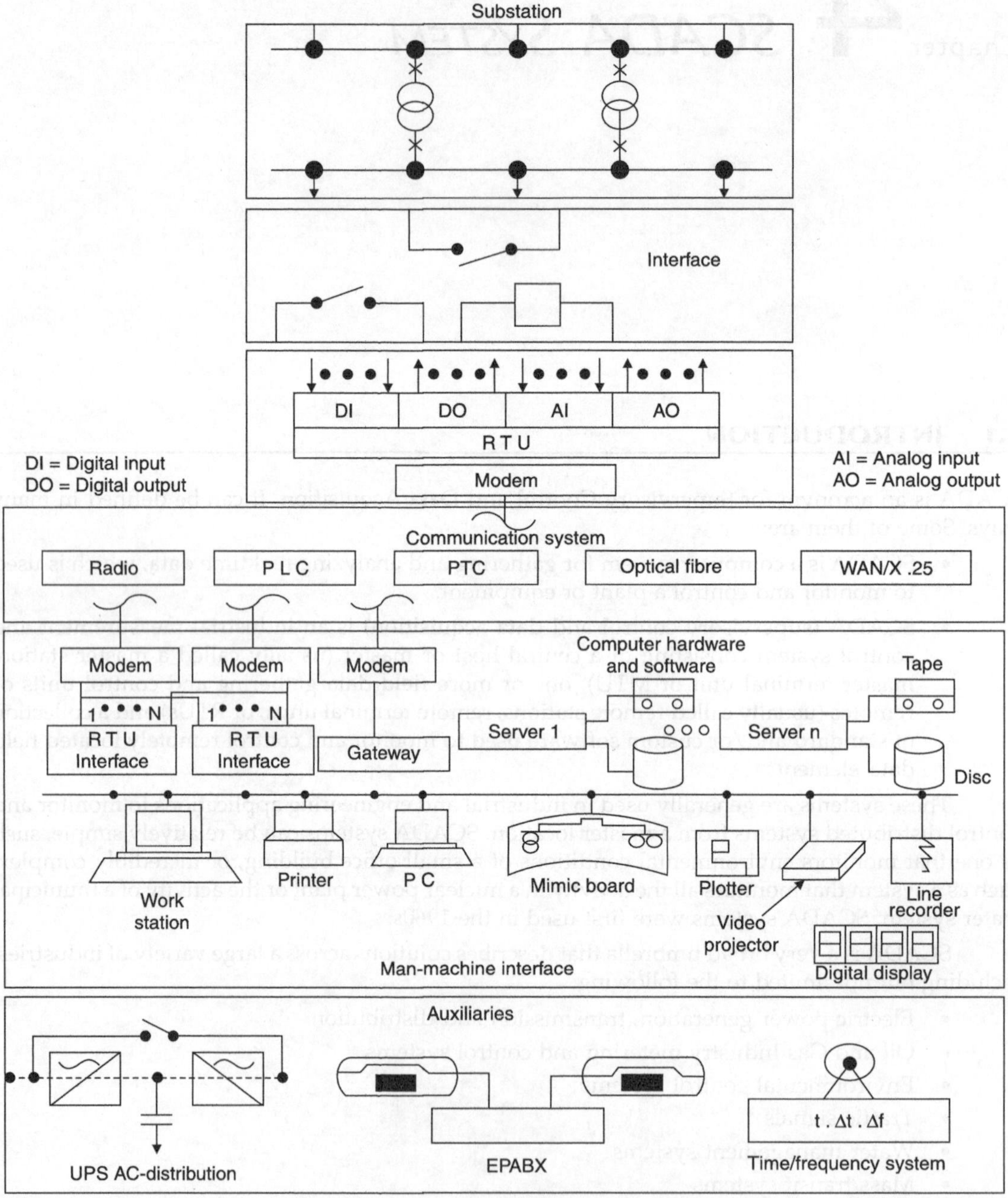

FIGURE 4.1 **Block Diagram of a typical SCADA system**

A typical arrangement of SCADA and other related equipment in a system are shown in Fig. 4.2. It shows all the related components and the interconnections.

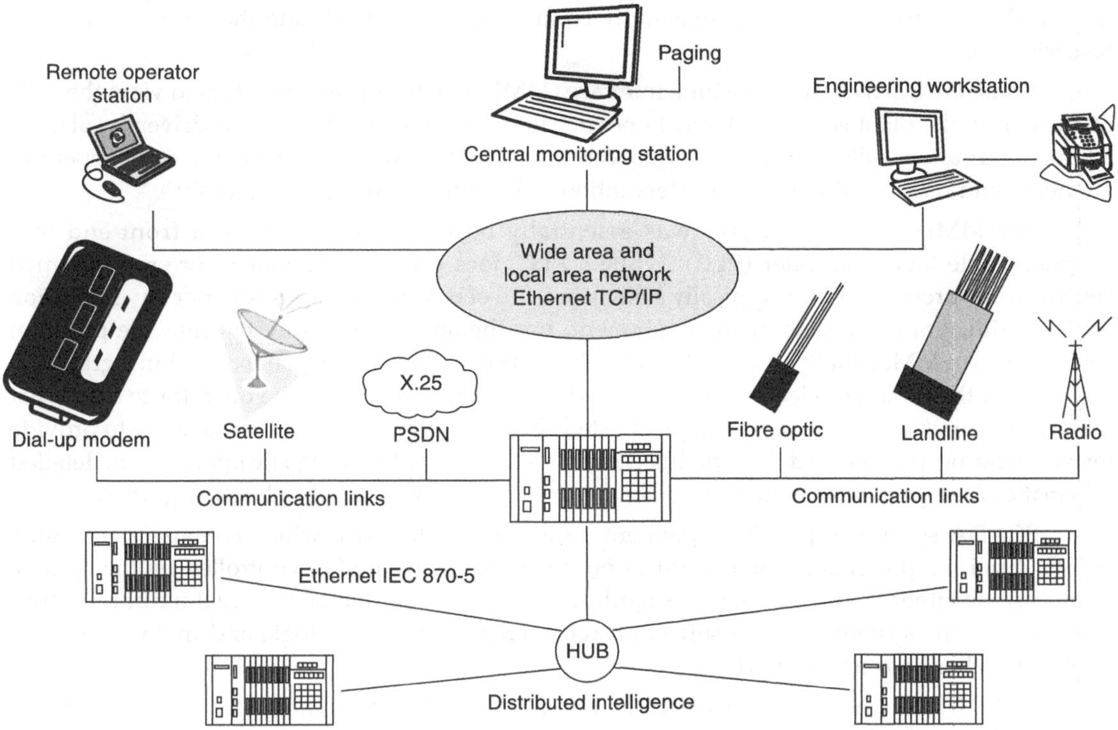

FIGURE 4.2 **A typical arrangement of SCADA and other equipment in a distributed system**

4.3 COMPONENTS OF SCADA

There are three components in a SCADA system, which are:

1. Multiple field Remote Terminal Units (RTUs)
2. Central Control Room with Host Computers
3. Communication infrastructure

4.3.1 Remote Terminal Unit (RTU)

The Remote Terminal Unit (RTU) connects to physical equipment such as switches, pumps, and other devices and monitors and controls these devices. As the term SCADA implies, the host computers allow for "supervisory level" control of the remote site and also "acquire data" from the remote field RTUs. The bulk of the site control is actually performed automatically by the RTU via a local field computer terminal. Host control functions are restricted to basic site over-ride or supervisory level capability.

Data acquisition begins at the RTU level and includes meter readings and equipment statuses that are copied or transferred to the host as required in a digestible format so that host control room operators can make appropriate supervisory decisions that may be required to over-ride normal RTU controls. Due to the sensitivity of the acquired critical data the entire system has restricted access.

While the SCADA man-machine interface (MMI) usually allows operators to view the state of any part of the plant equipment, most operator interaction with the system is driven by alarms. Alarms are automatically detected abnormal conditions in the plant equipment that require operator attention, and may require operator intervention to keep things running smoothly.

The MMI/SCADA industry was essentially born out of a need for a front-end to a programmable logic controller (PLC). While a PLC does provide automated, pre-programmed control over a process, a PLC is typically a blank box full of devices and does not offer any indication of the health, status, or state of the equipment, nor the ability to readily tap into the program commands. An MMI usually displays sensor information in its physical context, within a graphical depiction of the piping or electrical system in which it resides, allowing the operator to "see what the PLC is doing" to some extent. A sophisticated MMI may also be linked to a database to provide instant trending, diagnostic data, scheduled maintenance procedures, logistic information, detailed schematics for a particular sensor or machine, and expert-system troubleshooting guides.

SCADA systems typically implement a distributed database which contains data called points. A point represents a single input or output value monitored or controlled by the system. Points can be either "hard" or "soft". A hard point is representative of an actual input or output connected to the system, while a soft point represents the result of logic and math operations applied to other hard and soft points.

SCADA solutions often have Distributed Control System (DCS) components. Use of "smart" RTUs or PLCs, which are capable of autonomously executing simple logic processes without involving the master computer, is increasing. SCADA systems have traditionally used combinations of radio and direct serial or modem connections to meet communication requirements, although Ethernet and IP over SONET is also frequently used at large sites such as railways and power stations.

4.3.2 Central Control Room with Host Computers

This is the place where all the computational processes take place and the control signals are generated from the acquired data. For every SCADA system there is a host computer which performs all the required computational tasks as according to the requirement of the system for which it is working. The host computer is the main element of the central control room which is responsible for all control actions taking place in the whole system. All data acquired from the whole system is collected and sent into the control room for the processing to be done. After the data is processed the central control room completes the task of issuing control signals to the required part of the system as according to the result of the host computer.

4.3.3 Communication Infrastructure and Technologies

The reader may refer to Chapter 15 for an exhaustive coverage of communication technologies.

4.4 FUNCTIONS OF SCADA

Certain functions are basic to electric utility SCADA systems. The more common functions include:

- Data Acquisition
- Information Display
- Supervisory Control
- Alarm Processing
- Information Storage and Reports
- Sequence of Events Acquisition
- Data Calculations
- Special RTU Processing Control.

4.4.1 Data Acquisition

The periodic acquisition of data from RTUs is fundamental. Most systems for electric utilities acquire data on a polled basis where data are transmitted from the RTU to the master station only on receipt of a request from the master to the RTU. Two options are available on how the RTU responds. One option is to send the actual value or status of the point or group of points requested. The other option is to send only those points or groups of points where changes of state have occurred or where a value has exceeded a predefined delta change which has occurred since the previous poll request. This latter option is referred to as "reporting by exception". The main advantage of report by exception is a reduction in processing overhead at the master station. Average loading on the communication circuits is also reduced over the first option. However, sufficient communication circuit bandwidth capacity should be provided to accommodate the worst case situation when a large percentage of the points are rapidly changing, as when a major electric system disturbance occurs and the dispatcher's need for timely and accurate data is highest. Analog value report by exception is much less in vogue than discrete point report by exception.

The process of data acquisition can be considered as the collective process of several specialized and highly related sub processes. These sub processes include:

- Internal scanning and rapid update of the RTU internal database
- Periodic RTU polling by the master station
- Transmission of request data sets by the RTU to the master station
- Checking of the data for transmission induced errors
- Scaling of the data into engineering units
- Write over of the previous status or value in the database.

4.4.2 Information Display

Information display is the process of selectively retrieving both the fixed and the real-time data from the database, combining, and presenting them to the operator, usually in the form of limited graphics CRT colour pages. The fixed data include station online schematic information and other displayable information that is not varying in time. The variable data include status of two or three state devices, and analog values which vary in magnitude and possibly sign.

Display selection is most often organized in a hierarchical tree structure where index pages permit the operator to select a wide variety of displays using cursor positioning techniques including positioning by keyboard, trackball, light pen, or touch-screen methods.

4.4.3 Supervisory Control

Supervisory control is the process of actuating equipment operation at remote locations. The process includes selection of the station, selection of the device to be controlled, and execution of the desired command such as Trip or close. Correct selection and operation is critical for the safety of personnel and the security of the electric system. For this reason some form of select-verify selection-operate sequence or for short, "check-before-operate" method is employed. Operation of a device other than the selected device, or inadvertent operation of a device when no command has been made, must be avoided and is a critical design parameter for SCADA systems.

A variation of discrete supervisory control is "set point control." This form of control provides an operator or computer entered value as an analog output or an internal RTU value for proportional setting of an external device or control loop.

4.4.4 Alarm Processing

Alerting the operator to unscheduled events and informing him of the time of occurrence, the station location, the device ID, and the nature of the event is generally referred to as alarm processing. Many variations of alarm processing and presentation exist, the specifics being largely a function of the individual SCADA manufacturers. The most common output of the alarm process is chronological CRT alarm lists, hardcopy printouts, and audible alarms. Some recent systems provide synthesized voice alarms. Some form of alarm acknowledgment by the operator is nearly universal. In larger systems, significant human factor benefits can be realized by prioritizing alarms and by directing alarms only to operating positions concerned with a particular category of alarms.

4.4.5 Information Storage and Reports

Record keeping has always been an important task in the operation of electric systems. Accurate records are necessary to satisfy legal and governmental requirements, for accounting purposes, for support and forecasting of future system operations, and for engineering planning purposes.

The common practice for record-keeping is to capture pre-selected data sets at periodic intervals and save them in a rotating file. The periodicity of storage is frequently set at one hour intervals but specific electric system requirements may require more frequent capture and saves. The common time span for a rotating historical file is of the order of 40 days but spans of up to 12 months are common. The 40 day file provides one full month's data which can be analyzed and/ or archived to a bulk medium such as magnetic tape. Information older than the file time span is discarded.

The SCADA historical file provides an excellent source of information for the production of various reports. These reports, their formats and information content, may be customized to satisfy the specific needs of a wide variety of end users.

4.4.6 Sequence of Events Acquisition

Sequence of Events (SOE) is the process of capturing and recording unscheduled events with a time resolution between events of a few milliseconds. Historically, SOE recording has been provided with equipment dedicated to that purpose. Events are generally considered as discrete (two state) occurrences such as that with protective relaying systems. Oscillographic equipment has also been used for many years for recording transient analog values such as line current or voltage.

In recent years, SOE recording of discrete state changes has migrated into some SCADA systems. It is the process of capturing and recording unscheduled events to a time resolution independent of the periodicity of RTU polling by the central master station. An accurate time base or clock within the RTU is fundamental for SOE recording. Time resolution of events captured by a single RTU is typically ± 2ms. Accurate synchronization of multiple time bases located in a large number of widely separated RTUs presents a technical challenge. The most common approach is to establish RTU clock synchronization via the interrogate message. Time resolution between events captured by separate RTUs may be of the order of ± 8ms. The RTU time clocks are periodically synchronized by transmission of a time code message from the master station using the master-RTU communication channel with adjustments for the propagation time delay. The sequence of events at a given RTU may be printed locally at the RTU or may be transmitted to the master station where they are chronologically ordered with events received from other locations and then printed as one combined listing.

4.4.7 Data Calculations

A recurring need in SCADA systems is to perform various calculations utilizing the acquired data. Examples of data calculations with single variables are determination of averages, maximum or minimum values over given intervals, and integration with respect to time. Calculations with multiple variables may include sums, differences, products, quotients, squares, square roots, exponentiation, and others. A good practical application of such calculations is load monitoring of large transformer banks. The maximum capacity of large transformers is determined by their maximum allowable heat rise and transformer heating is closely associated with the MVA load. For various reasons, the most common metering associated with transformers is MW and MVAR but not amperes. MVA is then calculated periodically by the SCADA using the relation $MVA^2 = (MW)^2 + (MVAR)^2$ and placed in the real-time database where it is tested for limit excession. Boolean calculations may also be applied to discrete status points and the result considered as a new information point. This process has been referred to in some literature as"combinatorial processing". This process can determine a particular state of some part of the electric system not definable by only one status indication. A good example of combinatorial processing is the determination of whether a transmission line is energized or not. In most cases, circuit breakers at both ends of a line must be open for the line to be de-energized. Some lines have multiple circuit breakers at each end of the line (breaker-and-half scheme) all of which must be open to ensure the line is not energized from one end or the other.

4.5 SCADA APPLIED TO DISTRIBUTION AUTOMATION

4.5.1 Distribution Automation (DA)

Distribution Automation is the ability to remotely monitor and control the distribution network, collect information, and provide information in a useful manner to the users. Distribution Automation provides a building block for protection, control and monitoring of the distribution system. The definition for feeder automation varies from utility to utility. Some utilities refer to distribution automation as feeder automation, while others may refer to it as substation automation. Distribution Automation products are designed for interoperability and rapid automation implementation.

The DA system provides automatic reclosure of relays, automatic feeder switching and provides remote monitoring and controlling of distribution equipment such as transformers, capacitors, breakers, sectionalizes, communication nodes etc., from substation to Power System equipment and including consumer interface.

4.5.2 Why Integrate DA with SCADA?

The DA applications were to involve both monitoring and SCADA control functions. A major thrust of the DA project was focused on the methodology of data acquisition from the field devices and "seamlessly" transporting the information into the existing SCADA/EMS. The substation automation application being a third-party application, an integrated approach was intended to help the SCADA operator by providing an interface to carry out remote monitoring and control of some non-EMS applications. The request was also to give multiple users, including Planning and Engineering departments who currently have SCADA/EMS display monitoring capabilities, the ability to "view" the system to monitor power quality parameters apart from the Operations personnel.

There are certain other keen advantages to a total integration of the automation application with the SCADA system. The SCADA Operations personnel considered it cumbersome to have multiple pieces of hardware (in the form of stand-alone operator interface monitors or even PCs) to clutter up their desks. The ability to "open" the automation application windows in the SCADA console was the best option. This means that instead of handling a number of analog devices individually, now they can be handled and controlled from a single computer terminal.

4.6 ADVANTAGES OF DA THROUGH SCADA

There are many advantages of implementing DA through SCADA. Some of the main advantages can be given as:

4.6.1 Centralized Control

In a SCADA system the whole controlling action takes place in the Central Control room and the host computer is responsible for all the control actions taking place in the system. As all the data of the whole system gets processed at a single place the possibility of erratic processing of the data is minimized. Thus, it ensures the issue of correct controlling signals, which becomes the major advantage of SCADA when used in DA.

4.6.2 Economically Advisable

It is a true fact that the initial cost of employing a SCADA system is high, but it offers the advantage of good revenue in terms of minimizing the running cost of the control system for the whole system. Furthermore, SCADA can also help in deferring the capital cost for the new control equipment which is required by the conventional system.

4.6.3 Scope for Further Expansion

One of the main advantages of employing a SCADA system in DA is that it can initially be implemented for a small part of the system and can then be expanded for the rest of the system in steps.

4.6.4 Highly Reliable Information Storage

SCADA offers highly reliable information storage of the previous data of the system. This feature can be efficiently used for the purpose of forecasting required for system. Thus, the storage of information is a very important advantage for the SCADA in DA.

4.6.5 Other Advantages

1. Increased reliability of power for the consumers
2. Lower operating costs for the utility
3. Forecast of accurate demand and supply management
4. Faster restoration of power in case of a downturn
5. Quick alternative arrangement for power for important or emergency locations
6. Reduces human influence and errors
7. Offers complete power system controls and data acquisitions in a central locations
8. Ensures reliability and quality of power supply (minimum fluctuation in voltage supplies) for the consumers

4.7 REQUIREMENTS AND FEASIBILITY

4.7.1 Expected Functional Features

Real-time trending and phasor plots of electrical parameters such as current and voltage were expected to be part of the basic offering of the DA application package. It was also felt necessary for the application software to archive the acquired data (such as switch-operation data) in a format that other applications, such as geographical information system (GIS), would be able to use for study and analysis (*e.g.*, trouble call analysis function). Data archiving and historical reporting features were important considerations in choosing a vendor. From a system configuration perspective, the intention was to implement intelligent processing at the distribution pole level. An overburdened SCADA system is an evidence for erratic front-end processor (FEP) operations, maximum CPU utilization, and the field I/O capabilities being stretched to their limits. The intelligent controllers to be used for distributed processing were expected to relieve the SCADA host processor of the extra burden of DA monitoring and control. It was also specified at the very beginning that the monitoring and control equipment installed in a substation local area network (LAN) environment and communicating with the SCADA host processor, would have to adhere to industry-standard communication protocols. The hardware interface and the communication protocol must be able to interface with other vendors' intelligent electronic programmable controllers (IEPCs) and intelligent electronic devices (IEDs), such as logic controllers, smart meters, relays, etc.

In addition to the mandatory system performance needs, the automation system of choice was to be flexible enough for future system expansions and application enhancements by providing the necessary application programming interfaces (APIs). Nonproprietary system architecture was of paramount importance.

4.7.2 Selection Criteria

The first step in the quest for such an automated system was to identify the areas that needed the most re-engineering. In addition to monitoring of automated switches and capacitor bank switch control, future applications (such as fault detection and isolation and load management and control) were kept in mind during the system specification stage. GP & L system engineers based the basic selection criteria of the intelligent automation equipment on three simple factors:

1. Data acquisition scheme and control mechanism
2. Usability of the system
3. Cost.

All of these factors were given equal importance during the selection process. Several different sub- factors were considered under each category. And the overall emphasis was placed on "seamless integration" with the existing SCADA/EMS.

4.8 DA INTEGRATION MECHANISMS

There are a couple of scenarios that a utility may find itself in before trying to implement a DA application. In the first case, the utility is trying to acquire a DA system along with a new SCADA system. In the second case, the utility may already have a SCADA/EMS, and the substation automation application is being considered as an add-on. Although most of the integration issues are still going to be basically the same in both cases, there are certain differences that are worth mentioning.

In the first case, the utility may be in a better position from an implementation point of view when it is acquiring both the SCADA and the DA systems together. It is even conceivable that both products may be purchased from the same vendor. If not, then it is easier to work outmost of the potential software and hardware interface incompatibilities during the system specification stage.

In the second case, the situation may be a little more complex. The following criteria should be considered when attempting to integrate the DA and SCADA processes. The communication link and the protocol and data conversion mechanism is an important factor for the automation system intelligent electronic devices (IEDs), such as logic controllers, smart meters, relays, etc. In addition to the mandatory system performance needs, the automation system of choice was to tie into the SCADA/EMS host computer. There are several ways to implement a substation automation scheme. It is the responsibility of the system designer to choose the automation tool that will best suit the application. This discussion involves the selection of data acquisition and control device to be located in the substations and/or at the distribution poles. These devices may be programmable logic controllers (PLCs), intelligent remote terminal units (IRTUs), or even simple circuit monitors (CMs). The actual selection depends on several factors. For example, the following are three methods of substation automation implementation with varying degrees of complexity:

- Simple CMs may be used for monitoring purposes. To include control operations using CMs, 1/0 units would need to be added. The CMs communicate with a DA master station, which may be a stand-alone system or may be integrated with the SCADA system.
- The existing RTU infrastructure of the SCADA system may be utilized. Protocol conversion and data transport into the SCADA system **is** less complex in this case. But one has to determine if the existing RTU has enough communication ports available. The RTU in this type of setup is essentially being used as the data concentrator.
- More complex among the possible options is a direct hardware and software integration of the DA application with the existing SCADA system.

The size of the network can increase because of the increase in the number of data acquisition points, and/or due to the increase in the number of substation LANs bringing in the data to the master station. Integrability of the automation IEPC (be it an IRTU or a PLC) of choice in a substation LAN environment becomes a prime issue. The need for *peer-to-peer* vs. *master-slave* configuration between the IEPCs has to be determined. It must be determined if the LAN is deterministic or nondeterministic. If the DA application is for a protection scheme, it is imperative that the IEPC processes data in a pre-emptive fashion if there is any possibility of time-based operations being compromised. Cost per 1/0 point is one of the more important driving factors in the system design process. An IEPC may have a low cost for low 1/0 count. But it must be determined if cost stays low for higher 1/0 count. Some automation tools (typically the generic PLCs) maybe available off-the-shelf, thereby being less costly initially. But a cost analysis with all the SCADA integration and electric utility specific features must be done to verify its cost effectiveness against a comparable customized RTU. Standards-based communication protocol helps to keep the system integration costs down. System reliability, availability, programming features, and ease of use by the operator are important issues that must be taken into account. The DA application software becomes a key factor also **as** it is the driving force behind effective data transport to and from the SCADA system and productive presentation of information to the system operator via a graphical user interface (GUI). The applicability of these schemes would again depend on the particular state of automation and the existing infrastructure of the utility.

4.9 COMMUNICATION PROTOCOLS IN SCADA SYSTEMS

4.9.1 Definition

Protocol is a specification of an interface between modules running on different machines, as well as the communication service that those modules implement. The term is also used to refer to an implementation of the module that meets this specification.

4.9.2 Properties of Transport Protocol can be Expected to Provide

- Guarantees message delivery
- Delivers messages in the same order they are sent
- Delivers at most one copy of each message
- Supports arbitrarily large message
- Supports synchronization between the sender and receiver
- Allows the receiver to apply flow control to the sender
- Supports multiple application process on each host.

4.9.3 Types of Protocols

4.9.3.1 User Datagram Protocol (UDP)

The simplest possible transport protocol is one that extends the host-host delivery service of underlying network into process-to-process communication service. Transport protocol of the Internet architecture that provides a connection less datagram service to application level process.

4.9.3.2 Transport Control Protocol (TCP)

In contrast to simple protocol like UDP, a more sophisticated transport protocol is one that offers a reliable, connection oriented, byte-stream service. Such a service has proven useful to a wide assortment of applications because it frees the application from having to worry about missing recorded data.

4.9.3.3 Remote Procedure Call (RPC)

A common pattern of communication used by the application programs is the request/reply paradigm, also called message transaction. TCP does not match the request/reply paradigm very well either since going to the trouble of establishing a TCP connection just to exchange a pair of messages seems like overkill. So a new protocol *i.e.*, RPC-that more closely matches needs of an application involved in a request/reply message exchange.

There are two main problems that make RPC more complicated than local procedure calls.

- The network between the calling process and the called process has much more complex properties than backplane of a computer. For example, it is likely to limit message sizes and has a tendency to lose and reorder messages.

- The computers on which the calling and called processes run may have significantly different architectures and data representation formats.

Thus, a complete RPC mechanism actually involves two major components:

1. A protocol that manages the messages sent between the client and the server processes and that deals with the potentially undesirable properties of the underlying network.
2. Programming language and compiler support to package the arguments into a request message on the client machine and then to translate this message back into the arguments on the server machine, and likewise with the return value(this piece of the RPC mechanism is usually called a stub compiler)

4.9.4 Simple Network Management Protocol (SNMP)

Simple Network Management Protocol (SNMP) is the only protocol for element and network management that has achieved broad acceptance among many proprietary solutions. Originally it was designed only as a short-term solution until other protocols would be available. Due to its simplicity and low resource consumption at the managed device, it is still the protocol of choice for the majority of LAN devices to provide at least some status information. So far, three versions of SNMP have been defined. In the following, we will restrict our considerations to the most widely used first version SNMPv1 (RFC1155, RFC1157, RFC1213). For this protocol, lots of solutions —from simple command-line tools to fully integrated platforms (HP-Open View, Cabletron Spectrum, and IBM Tivoli)—have been developed. From today's view, even if SNMP might not be the best protocol for all situations, it is the one with the largest number of readily available tools and application programming interfaces (APIs).

SNMP defines a clear client/server relationship between devices. The *manager* communicates with several *agents* to retrieve data or set new values on the managed device. All accessible data at the agents are stored in a management information base (see description below). If a network element is not directly addressable by the management protocol, a *proxy agent* provides a conversion between the proprietary protocols of foreign agents and the SNMP environment. SNMP only describes *how to represent* management information and *how to transfer* it. It makes no suggestions about graphical representation, tools, or the like.

SNMP is an application-layer protocol, designed for using the connectionless service Used Datagram Protocol (UDP) as transport layer. Only five messages are defined for the communication between manager and agent (Figure 4.1): *GetRequest*, *Get- NextRequest*, and *SetRequest* are issued by the manager and answered with a *GetResponse* from the agent. The losses of a PDU are usually handled by the manager application with a timeout and retry mechanism. A *Trap* is sent by the agent (very much like an interrupt) to report an important situation to the manager which then has the possibility to immediately react upon this information. Especially in large, complex installations and with respect to latency and bandwidth consumption, this mechanism is superior to a polling approach. Both mechanisms can also be combined (trap directed polling) for further optimization. In competing approaches like HTTP, where communication is normally initiated by the client, the lack of an equivalent feature is typically circumvented by proprietary solutions such as sending notifications via SMTP.

All information at the agent is stored in a tree-like structure called a *management information base* (MIB). Its contents are described in the notation of ASN. All data that can be addressed and manipulated by management applications must be part of the MIB. Each element of the MIB is unambiguously identified by assigning it a globally unique *Object Identifier* (OID), a sequence of nonnegative integers delimited by periods. In addition, a textual name is associated with each OID. One important feature of the SNMP MIBs is that they are defined in a worldwide unique name space. Several subtrees of this name space are standardized MIBs, whereas others can be reserved, for example, by companies to add individual extensions. One of the most valuable aspects of SNMP is the plethora of MIBs containing thousands of OIDs already standardized and ready for use.

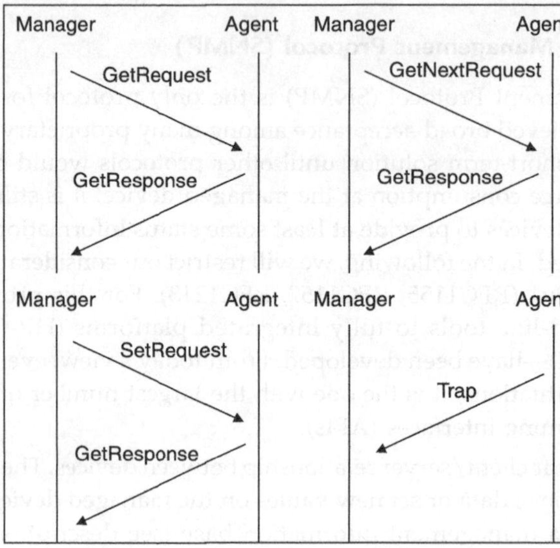

FIGURE 4.3 **SNMP protocol operations**

4.9.5 MODBUS Communication Protocol

According to the OSI model, the MODBUS communication protocol belongs to the data link layer. However, it is specified as a RS-232, RS-485 or Ethernet physical layer, working in a client-server architecture. On data link layer, the frame format has four fields: device address, function code, data and error check (CRC). Specifically, the device address is a byte field and it contains the destination device address. The function code is a byte field and it indicates whether it is a request or a control frame. The data field has a variable length and indicates two parameters. One of them is the number of bytes that will be read and the other one is the initial address register which has the data bytes that will be read. The CRC is a two byte field and implements the error verification algorithm. Nevertheless, each measurement taken by the PM650 meter has a logical address specified by the MODBUS protocol. According to this, for example, in the PM650, the address 1000 corresponds to the frequency value for the voltage and electrical current signal.

4.9.6 Migrating Electrical Power Network SCADA Systems to TCP/IP and Ethernet Networking

Traditional SCADA systems for electricity utilities rely on data transmission over fixed analogue circuits and modems. This method has been satisfactory over the years but it is becoming obsolete and unsuited to today's requirements. The use of TCP/IP technology can overcome the limitations of analogue communications, allow the network to be more flexible in terms of expansion and reconfiguration, and have higher bandwidth potential whilst retaining the qualities essential for SCADA operations.

4.9.7 What is Driving the Need for Change

Traditional **SCADA** networking is based on fixed voice grade circuits and modems thus guaranteeing that a communication path is there when required and that transmission delay and variations are very small. This technology is becoming obsolete and unsuited to the developing needs of toady power system operations. In recent years, telecommunication systems have undergone radical change prompted mainly by the desire to increase system performance. New technologies are providing huge increases in performance at lower and lower unit cost. At the same time, deregulation and privatizations of the electricity industry coupled with the liberalization of the telecommunication market have imposed new requirements on the electrical power communication network. Key drivers for change of traditional SCADA networking include the following:

- **Optical fibre installations:** Optical-fibre technology is regarded as the most powerful and (e-mail) versatile communication medium due to its high-bandwidth capability and immunity from electromagnetic interference. The use of optical fiber as a digital communication medium has, become commonplace in most electricity utilities as the benefits of implementing fibre-optic communication systems and decreasing costs have become more apparent.
- **TCP/IP networking:** Transmission control protocol/Internet protocol (TCP/IP) suite and associated technologies have become the *de* facto world standard for data networking. They are used to deliver the Internet and corporate intranets using connectionless networking. TCP/IP has not been ideally suited **to** the transport of SCADA traffic because transmission times can vary and performance under extreme conditions may not be predictable. However, a great deal of development work has taken place to produce quality of service (QoS) standards that will guarantee delivery of connection-oriented traffic over TCP/lP networks.

Emergence of *QoS* over TCP/IP: The technical obstacle preventing the implementation of SCADA on TCP/IP networks has been the absence of a standardized method for providing quality of service guarantees over TCP/IP data networks. QoS standards are central to the successful provision of TCP/IP **service** to networked applications that have differing requirements. For example, SCADA data traffic requires very low and consistent transmission delays but limited bandwidth administrative data and e-mails require higher bandwidth but are far less critical of time delays and variations. QoS technologies are designed to enable a network to provide transport services appropriate to the needs of each application.

4.9.8 Why Move SCADA Networking To TCP/IP

The TCP/IP networking protocol mite was originally devised to support military research projects in the USA. It went on to become the standard used by the Internet. Partly due to its simplicity and its rapid standardization procedures TCP/IP has become the de facto international standard. Conventional SCADA network designs rely on the predictable nature of connection oriented services using fixed audio bandwidth links, analogue modems and specific protocols. Setting up and maintaining this network calls for specialized shills. Reconfigurations involve hardware rewiring, are time consuming and costly Bandwidth is limited to 3 kHz, which is adequate for current RTUs hut potentially limiting as businesses move towards the use of substation automation and remote management. The support of analogue modems is becoming increasingly difficult. Their use is already limited and will diminish further as the world moves to digital communications. Changing to TCP/IP networking will enable the management of SCADA networks to be integrated into a system common to the corporate data network. Reconfigurations will he simplified to keyboard commands rather than rewiring at multiple points. Bandwidth can be allocated as required and RTUs themselves remotely managed.

The advantages of TCP/IP networking include:

- Worldwide adoption (*e.g.*, the Internet)
- Very well developed hardware and software market
- Simplicity and choice of application layer protocols
- Inherent resilience of the IP routing concept
- Strong network management, including remote control and monitoring.

Using TCP/IP and the commonly associated Ethernet technology will give power system operators access to a wide range of standards based inexpensive hardware and a large pool of trained staff.

TCP/IP has not been widely introduced into wide area SCADA communication systems to date. This is partly due to the replacement cycle but also because of the non-deterministic nature of Ethernet and TCP/IP communications. The development and implementation of QoS standards removes the risks associated with the non-deterministic nature of the underlying connectionless networking. As TCP/IP and Ethernet support is becoming available from SCADA equipment manufacturers, the use of wide area TCP/IP networking from master stations to RTUs is practicable. SCADA applications have historically used a very wide variety of application level protocols. Two standards are emerging which are both designed to operate successfully with TCP/IP and Ethernet.

LONG TYPE QUESTIONS

1. Define SCADA and its objectives.
2. Give examples of applications of SCADA.
3. What are the components of SCADA?
4. Write a short note on RTU and describe the technologies involved.
5. List the various functions of SCADA.

6. Explain the various methods of Data Acquisition.
7. What are the different sub-processes involved in Data Acquisition.
8. Write a short note on the following:
   (*a*) Supervisory control
   (*b*) Alarm Processing
   (*c*) Sequence of events acquisition
   (*d*) Data Calculations.
9. What is combinatorial processing?
10. Why should DA be integrated with SCADA? Explain its advantages.
11. What criteria should be considered while integrating DA with SCADA?
12. Define protocols and enumerate their types.
13. Define MIB and OID.

MULTIPLE CHOICE QUESTIONS

1. Which among these components connects to physical equipments of the distribution system
   (*a*) RTU (*b*) Central Room Computer
   (*c*) Communication Infrastructure (*d*) None of the above.
2. The function of data acquisition begins at
   (*a*) communication infrastructure (*b*) central room computer
   (*c*) communications infrastructure (*d*) none of the above.
3. In a SCADA system the operator interaction is driven by
   (*a*) Communication Infrastructure (*b*) PLC
   (*c*) Alarm (*d*) RTU.
4. In a SCADA system the host computer performs
   (*a*) Data Acquisition functions (*b*) Computational tasks
   (*c*) Communication (*d*) All of these.
5. Information Display is achieved in the form of
   (*a*) Limited Graphics (*b*) CRT Color Pages
   (*c*) Both (*a*) and (*b*) (*d*) None of the above.
6. During supervisory control which of the following methods is employed for verification
   (*a*) Set point control (*b*) Report by exception
   (*c*) Check before operate (*d*) Sequence of Events acquisition.
7. Alarm processing refers to which of the following
   (*a*) Acquiring data from the field RTUs (*b*) Set point control of the RTUs
   (*c*) Alerting the operator to unscheduled events
   (*d*) Retrieving real time data.

8. Accurate record keeping is essential for
   (*a*) Legal requirements (*b*) Accounting purposes
   (*c*) Forecasting purposes (*d*) All of the above.
9. Capturing and recording unscheduled events with a time resolution of a few milliseconds is referred to as
   (*a*) Information Storage (*b*) Supervisory Control
   (*c*) Data Calculation (*d*) None of the above.
10. The typical time resolution of events captured during a SOE operation would be
   (*a*) 5ms (*b*) 10 ms
   (*c*) 2 ms (*d*) 3 ms.
11. Calculation of average and integration wrt to time would be a part of
   (*a*) Information Display (*b*) Sequence of Events Acquisition
   (*c*) Data Calculations (*d*) Alarm Processing.
12. Determination of whether a transmission line is energized or not is a good example of
   (*a*) Sequence of Events (*b*) Information Display
   (*c*) Combinatorial Processing (*d*) Data Acquisition.
13. The highly reliable information storage that a SCADA system offers can be sued for
   (*a*) Distribution Automation (*b*) Minimising Cost
   (*c*) Centralised Control (*d*) Forecasting.
14. Which of the following protocols offers a byte stream service
   (*a*) UDP (*b*) TCP
   (*c*) RPC (*d*) SNMP.
15. Majority of LAN devices employ SNMP due to
   (*a*) It is cheap (*b*) It is simple
   (*c*) It has low resource consumption (*d*) Both (*b*) and (*c*).

REFERENCES

1. Peter Brennan, Ashley Weke and Robert Thornton; *Integration of SCADA and DA/DMS Across a Large Distribution System*; IEEE Catalogue No. 95TH8130; Pages 31–36
2. Dr. Peter Kadar, Dr. Kovacs Attila, Mohacsi Andras and Szabo Ervin; *Extension of the Standard SCADA Functionality with AI Tools*; Electric Power Engineering, 1999; PowerTech Budapest 99; International Conference on 29 Aug.-2 September 1999; Page 27
3. Samuel C. Sciacca and Wayne R. Block; *Advanced SCADA Concepts*; ISSN 08950156/95/ $4.0001995 IEEE; January 1995; pages 23–28
4. Koustuv Ghoshalm; *Distribution Automation: SCADA Integration is Key*; ISSN 08954156/97/ $10.00019971EEE; January 1997; pages 31–35
5. Charles H. Neeley, Patrick W. Abshire and Daniel J. Hoge; *Programmable Controllers Integrate oil field SCADA and Automation*; 1990 IEEE; pages 235–241

6. Dennis J. Gaushell, Henry T. Darlington; *Supervisory Control and Data Acquisition*; Proceedings of the IEEE, Vol. 75, No. 12, December 1987; pages 1645–1658
7. A.S.Pabla; *Electric Power Distribution*; 4th edition; TATA McGraw Hill
8. A.S.Pabla; *Electric Power Distribution*; 5th edition; TATA McGraw Hill
9. http://www.webopedia.com
10. http://www.sss-mag.com
11. http://www.arcweb.com
12. Larry L. Peterson and Bruce S.Davie, "*Computer Networks A Systems Approach*", Elsevier, 3rd edition, 2005.
13. Michael Kunes and Thilo Sauter "*Fieldbus-Internet Connectivity: The SNMP Approach*" IEEE TRANSACTIONS ON INDUSTRIAL ELECTRONICS, VOL. 48, NO. 6, DECEMBER 2001.
14. T. Sauter, M. Lobashov, G. Pratl. "*Lessons Learnt From Internet Access to Fieldbus Gateways*"; IECON'02 The 28th Annual Conference of the IEEE Industrial Electronics Society, Sevilla; Nov. 2002.
15. F. H. Ramírez Leyva, J. P. Acevedo Cuellar, R. G. Maldonado Basilio, E. Espinosa Justo "*Wireless System for Electrical Networks Testing Based on MODBUS Protocol*" Proceedings of the 14th International Conference on Electronics, Communications and Computers (CONIELECOMP'04).
16. Kwok-Hong Mark and Barry Holland, "*Migrating electrical power network SCADA system to TCP/IP and Ethernet networking*", Power Engineering Journal, December 2002. Pages 305 – 311.
17. John Gillerman, Herb Falk, and Ralph Mackiewicz "*Focus on IEC TC 57 power system reliability and profitability*", IEEE Power & Magazine, July/august 2005, pages 66–67.
18. http://www.ieee.org/power/iectc57.pdf.

Chapter 5 SUBSTATION AUTOMATION

5.1 INTRODUCTION

In today's deregulated marketplace, utilities are expected to provide reliable power at a competitive price. Automated substations can provide the information needed to maintain uninterrupted power to the customer at a lower maintenance cost. Substation automation is the integration of smart electrical equipment (*e.g.*, circuit breakers, transformers, relays, etc.) that has the ability to monitor their functionality. For example, circuit breakers are able to measure their contact resistance and acknowledge proper maintenance diagnostics. Automated substations consist of microprocessor-based relays, circuit breakers, transformers, and motor-operated air switches that are monitored by a graphical interface unit that can be remotely accessed.

Substation automation is not a new concept. Substations have been equipped to perform automatic re-closing, automatic bus sectionalizing, automatic load transfers, automatic capacitor switching, etc. for many years. In the past, however, these and other functions were implemented using a combination of control panels, auxiliary relays, switches, lights, meters, transducers and extensive wiring and cabling. In many applications today, however, substation automation is perceived as being something "new". This perception is probably because developments in substation equipment have expanded the potential capabilities of substation automation far beyond that which could previously be reasonably accomplished. The principal development is generically defined as an Intelligent Electronic Device (IED) which typically consists of one or more microprocessors and communications ports; with the ability to transmit data and execute control commands, and frequently provide a local user interface. Typical examples are relays, meters, and specialized sensors. Prior to the introduction of microprocessor-based relays, the protection and control of a very small substation consisting of one incoming line, one transformer and two feeders would require four large panels filled with relays, switches and lights. Only one panel is required when microprocessor relays are used. Interestingly, at the same time the space requirements are reduced by a factor of four, so is the installation cost.

5.2 DEFINITION OF SUBSTATION AUTOMATION

5.2.1 Meaning of Substation Automation

- To obtain supervisory control and data acquisition (SCADA) data collection and control capabilities at the substation. The substation's function and operation can be monitored and controlled from a remote location. Communication can be made via modem, high-speed communication lines, and fibre optic cables.
- To use microprocessor-based relays to perform online monitoring and control operations. This eliminates having to build a separate SCADA network to accomplish this task, which eliminates redundant sensors, wiring, and transducers.
- To convert the traditional switchboard look (*i.e.*, panels, mechanical switches, red and green lights, etc.) to a graphical user interface (GUI) accessible through the station PC.

This time-based data will be used to perform specialized functions and provide decision-making guidance to operators, maintenance planners, and engineering.

5.2.2 What Makes Substation Automation "New"

Advances in communications technology are used to tie everything together into a useful network. Within the substation, a single high-speed Local Area Network (LAN) is used to transmit data and control commands, replacing the extensive and costly cables that had been required.

Presently, a number of different LAN techniques and protocols are in use. The industry is actively working on development of a new standard LAN definition that will be based on the use of Ethernet and Manufacturing Messaging Specification (M M S) and will be compatible with the Utility Communications Architecture (UCA). There are already many techniques for moving data out of the substation to a master station or to other substations. These include the use of leased or dedicated telephone lines, dial-up phone lines, cellular telemetry techniques, satellite transmissions, various flavours of radio techniques and fibre-optic networks. Basically, this variety of communications methods results in the ability to transmit large amounts of information at a rapidly declining cost per bit. The combination of microprocessor-based devices and communications technology creates the ability to obtain more information about the power system and the equipment being used. Power system variables include magnitude and angle of voltages and currents, real and reactive power, frequency, power factors etc. Information is available regarding the initiating event for relay operation, the location of faults, and fault analysis. Specialized sensors and transducers are used to build a database relating to equipment condition and use; so that analysis techniques can be used to determine equipment condition and or base maintenance activities on actual condition rather than time schedules. Within the substation, the use of Programmable Logic Controllers or other types of computers opens up a vast array of automation possibilities. Complex schemes for dead bus and dead line reclosing can be implemented, with the sequence being based on actual power system conditions that exist at the time. Re closing of circuits can be modified based on cold load pickup requirements. Load transfers between busses and transformers can be made to protect against transformer overloads. Bus voltages and power factors can be tightly controlled to minimize losses or voltage variations. Supplementary measurements and inputs can be used to initiate automatic equipment re-energisation after a transformer or bus differential.

5.3 WHAT IS DRIVING SUBSTATION AUTOMATION

- Deregulation of the utility industry is creating a situation where utilities must automate and acquire more information in order to remain competitive. Reduced costs of installing, operating and maintaining the utility plant is a major objective. There are significant and quantifiable cost savings that can be achieved through the use of substation automation. Somewhat less tangible objectives include improved efficiency, improved system reliability, improved power quality, and improved customer service. All of these items, in one or more ways, will be important in customer retention.
- "Open Access", sometimes referred to as wholesale and retail wheeling creates a number of requirements for additional information about the power system. The objective of open access can be said to be making maximum utilization of the capabilities of the electric power system, without violating any stability or equipment operation limits. Some states are mandating the use of Independent System Operators (ISO) to control the operation and use of the power system in a safe and reliable manner. The ISO uses computers and complex software to perform real-time calculations of transmission capacity, stability, contingency identification and evaluation, and operating costs. All of these programs, however, are based on the use of power system data from the substation. It is self-evident that if the quality and quantity of information available is inadequate, then the results of the calculations will be less precise, thereby requiring greater "safety margins" to be applied. It can be expected that the ISO will demand more and better data in order to insure that maximum benefits of open access are obtained without creating undue risk of blackouts or other system operating problems.

5.3.1 Benefits of Substation Automation

Electrical utilities are just as cost-conscious as any other corporation today. Increased competition, decreased margins and regulatory changes demand efficiency at every level of the business.

By implementing substation automation we can:

- Save huge money on pricey standalone devices by taking advantage of integrated system functions that eliminate redundant and space-consuming equipment. That means only one set of hardware, one set of wiring, one communications system and one system to commission, operate and maintain.
- Eliminate site visits and cut operations and maintenance costs by remotely accessing and programming devices from a central station.
- Extend primary equipment life by collecting accurate and relevant operating data for just-in-time maintenance.
- Cut engineering and planning costs by sharing pre-processed data and analytical reports companywide with standardized file formats and common databases.
- Improve the quality of service with enhanced system reliability and safety.
- Reduce the frequency of outages by identifying problems before equipment fails.
- Restore service faster by pinpointing faults and improving response times with better alarms, indications and diagnostics.

5.4 FUNCTIONS OF SUBSTATION AUTOMATION SYSTEM

- Today, a major function of the Substation Automation (SA) system is to provide SCADA RTU functionality. Essentially, this means concentrating the data from the substation IEDs into one or more messages for transmission to the master station. Often, this can include protocol translation when collecting data from different IEDs. Typically, the SA System will include a local user interface for data monitoring and control; and a local database of real time and other data. Some systems include complete alarm and event processing. Some SA systems will also provide access and control for non-real-time information such as fault records, event records, oscillography, relay settings, load profiles, etc. The SA system will normally have a high-speed LAN to support local automation functions which are typically programmed into the SA system itself, or done using programmable logic controllers. To the extent possible, the LAN will provide peer-to-peer communications between devices in the substation. Once significant processing power in the form of a SA system is introduced into the substation, it becomes possible to support "beyond the fence" automation such as feeder and load transfers within the same substation or between substations; or feeder reconfiguration schemes that can minimize customer interruptions, improve power quality and achieve maximum utilization of installed equipment.
- The substation automation system includes functions that control devices in a substation based on data collected from devices in the same substation. The processing capability could be provided by a substation controller, a smart RTU, or the DMS computer.
- Substation automation functions include service restoration via bus sectionalizing, bus voltage control, substation parallel transformer circulating current control, line drop compensation, and automatic reclosing.

5.4.1 Automation Building Blocks

In a typical substation automation scheme, with SCADA integration, the basic components may consist of (but not limited to) the following:

- SCADA host computer
- SCADA front-end processor (FEP)
- Communication FEP to communicate with non standard LAN protocols
- Multiplexer with protocol converter
- Local user interface
- Remote user interface to reconfigure and troubleshoots
- Communications medium, such as fibre optic, radio, cellular, or copper wire
- Intelligent electronic programmable controllers (IEPCs)
- DA devices such as circuit monitors (CMs)
- Interfacing intelligent electronic devices (IEDs), *e.g.*, smart meters.

Fig. 5.1 shows power system data flow with integrated substation automation systems.

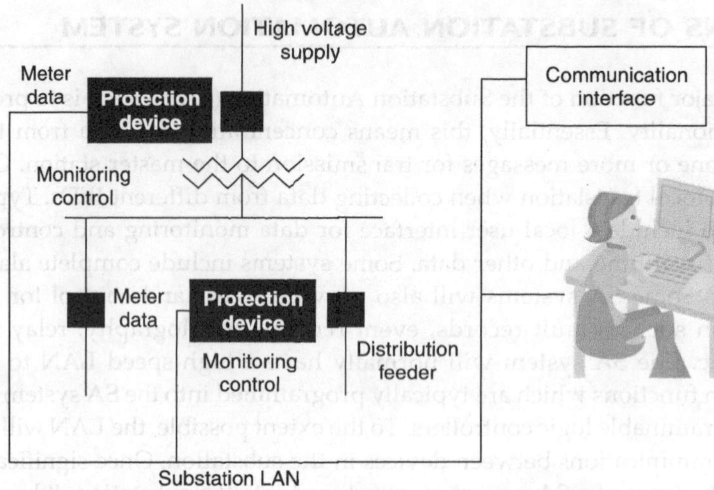

FIGURE 5.1 **Power system data flow with integrated substation—automation systems**

5.5 STATE AND TRENDS OF SUBSTATION AUTOMATION

Integrated protection and control first appeared in the mid eighties and has since then matured to full-scale substation automation.

5.5.1 Recent Practices

Simple systems for distribution or industrial networks mostly use feeder dedicated combined protection and control IEDs and a PC-based central unit. Alternatively, enhanced RTUs with

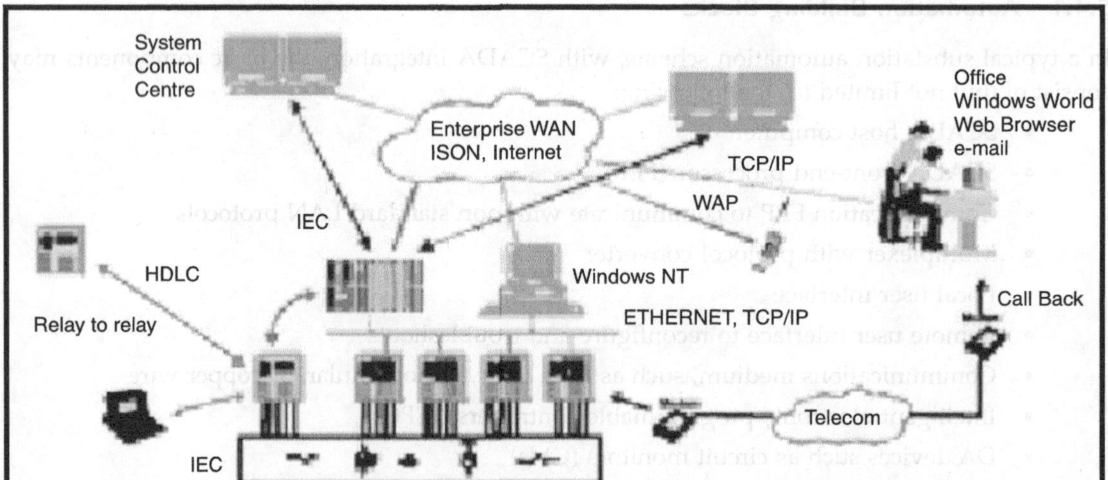

FIGURE 5.2 **Communication world of substation automation**

decentralized I/O peripherals are applied. Ethernet is generally accepted as substation LAN. Recently, Ethernet with TCP/IP has also been introduced. Larger systems typically use a special central unit (server) and separate bay units for control. The remote control function is emulated each in the central unit. Time accurate GPS based synchronizing is available as an option. Direct peer-to-peer communication between bay units is offered in some cases. It can be used for control (*e.g.*, interlocking), however not for protection because of the relatively long reaction time (some 100 ms). Figure 5.2 shows the complex communication world of protection and substation automation.

5.5.2 Internet Technology

The latest trend goes to using Internet technology in an Intranet or the Internet itself. Several vendors already offer substation automation systems with integrated Internet server. In this way, the acquired data can be exchanged in a cost saving way in an Intranet and distributed to a wider circle of users. Classic workstations can be replaced by normal Internet browsers. Maintenance work, for example implementation of new functions, must then only be performed at the central application server. In Japan, some systems have been in service where mini-servers are implemented in relays and bay control units on basis of Java VM (Java Virtual Machine). An American vendor is even a step ahead and offers a monitoring system where space and administration of data is provided on the vendor's own server. The user must only install the Internet enabled relays and devices in his substation and connect them to the Internet via the local service provider.

5.5.3 Highly Integrated Substations

The use of electronic sensors instead of traditional current and voltage transformers in combination with digital protection and control allows us to design compact substations. The current transformers are, in this case, designed as Rogowski coils or closed-core low-signal transformers. Resistive or capacitive voltage dividers are used as voltage sensors. The low signal level requires using shielded cables for the connection of the combined protection and controlling IEDs. This design approach considers the switchgear panel as one totally integrated module. In high voltage, the development goes to optic-electronic current and voltage transformers.

5.6 INTELLIGENT, AFFORDABLE SUBSTATION MONITORING AND CONTROL

An automated substation control and monitoring system protects high voltage equipment from faults, disturbances, and overloading on the power grid, and optimizes the transmission and distribution of power from the utility to its customers. For increased reliability, reduced fault restoration times, and balanced electrical demand with available generation, a host of substation equipment must be monitored and controlled: RTUs (remote terminal units), relays, breakers, reclosers, switches, capacitors, transformers, and more.

A traditional SCADA (Supervisory Control and Data Acquisition) system consists of a master station, RTUs, IEDs (intelligent electronic devices), transducers and sensors. The master station provides the primary operator interface and manages overall system functions, collecting and analyzing data from RTUs, as well as initiating control actions. IEDs, transducers, and sensors measure power, energy, voltage, current, and analog I/O values such as temperatures and pressures, forwarding data to RTUs. Traditional RTU-based SCADA systems are designed to meet the needs of large utilities that monitor tens of thousands of status points and require master stations that process thousands of alarms per second. Advanced applications tied to SCADA are required for managing their complex power networks. Very high power availability is critical, so these systems are fully redundant.

Traditional SCADA systems are expensive—cost per substation can exceed $100,000. Master station hardware and software are specially designed using expensive proprietary systems. The complexity of these systems requires additional investment for training and maintenance. Because the RTU is the central device in a substation and represents a single point of failure, many features have been added to improve its reliability. However, RTU enhancements are very expensive, further increasing long-term ownership costs.

Dedicated to providing mission-critical operational data, a traditional SCADA system has very limited ability to access and report on highly valuable non-operational data, such as power quality.

For utilities that need to cost-effectively share information between different departments, the inability of traditional SCADA systems to integrate operational, non-operational, and revenue metering data into one system is a significant disadvantage.

5.6.1 EEM Provides Intelligent, Affordable Substation Monitoring and Control

An enterprise energy management system provides intelligent, affordable substation automation monitoring and control for smaller utilities such as municipal utilities or rural cooperatives, and can provide advanced functionality for larger utilities when integrated with an existing SCADA system.

An EEM system is comprised of web-enabled software and intelligent revenue-class ION metering and control devices that calculate record and analyze hundreds of data points in real-time, store the information in onboard memory, and provide status monitoring. An ION meter can be programmed to perform automatic control, alarm on user-specified conditions, and offers advanced power quality analysis for troubleshooting.

Flexible communications options enable an EEM system to leverage existing infrastructures such as high-speed Ethernet and the Internet for reduced installation and data acquisition costs, as well as real-time data collection from geographically diverse sites. Easy integration with traditional SCADA systems allows EEM to provide value to legacy systems.

EEM technology forms a layer of "energy information technology" that correlates time synchronized, real-time data from intelligent devices throughout each substation and across the grid. An EEM system can send automated or event-driven reports inter-departmentally or to customers, and can be accessed using a standard web browser.

These capabilities enable EEM to provide an integrated substation automation solution that improves operations and maintenance, increases system and staff efficiencies, and minimizes capital investments.

5.6.1.1 Substation Automation Solution Using EEM

This section describes how smaller utilities can use an EEM system for integrated substation monitoring and control.

System Requirements:

A typical EEM substation automation solution might include the following elements:

At the control room:

- ION Enterprise operations software as the master control station
- Windows 2000 server

At the service entrance(s):

- ION 8500 – ANSI class 0.2 revenue meter, power quality meter and communication gateway to feeder meters
- ION 7600 – ANSI class 0.2 revenue meter, power quality meter and communications gateway to feeder meters

At the feeders:

- ION 7350 – real-time power, energy, voltage, power quality, and I/O monitoring

Substation equipment:

- RTUs, relays, breakers, reclosers, transfer switches, capacitor banks, transformers, insulators, and UPS/battery systems etc.

Communications:

- Ethernet Local and Wide Area Networks (LAN/WAN)
- Modem, radio modem, telephone lines
- Serial, twisted pair, RS-485/RS–232

Fig. 5.3 shows an EEM automated substation solution that consists of multiple remote substations connected to an ION Enterprise operations software master station using Ethernet, telephone modems, and spread spectrum radio links for communications.

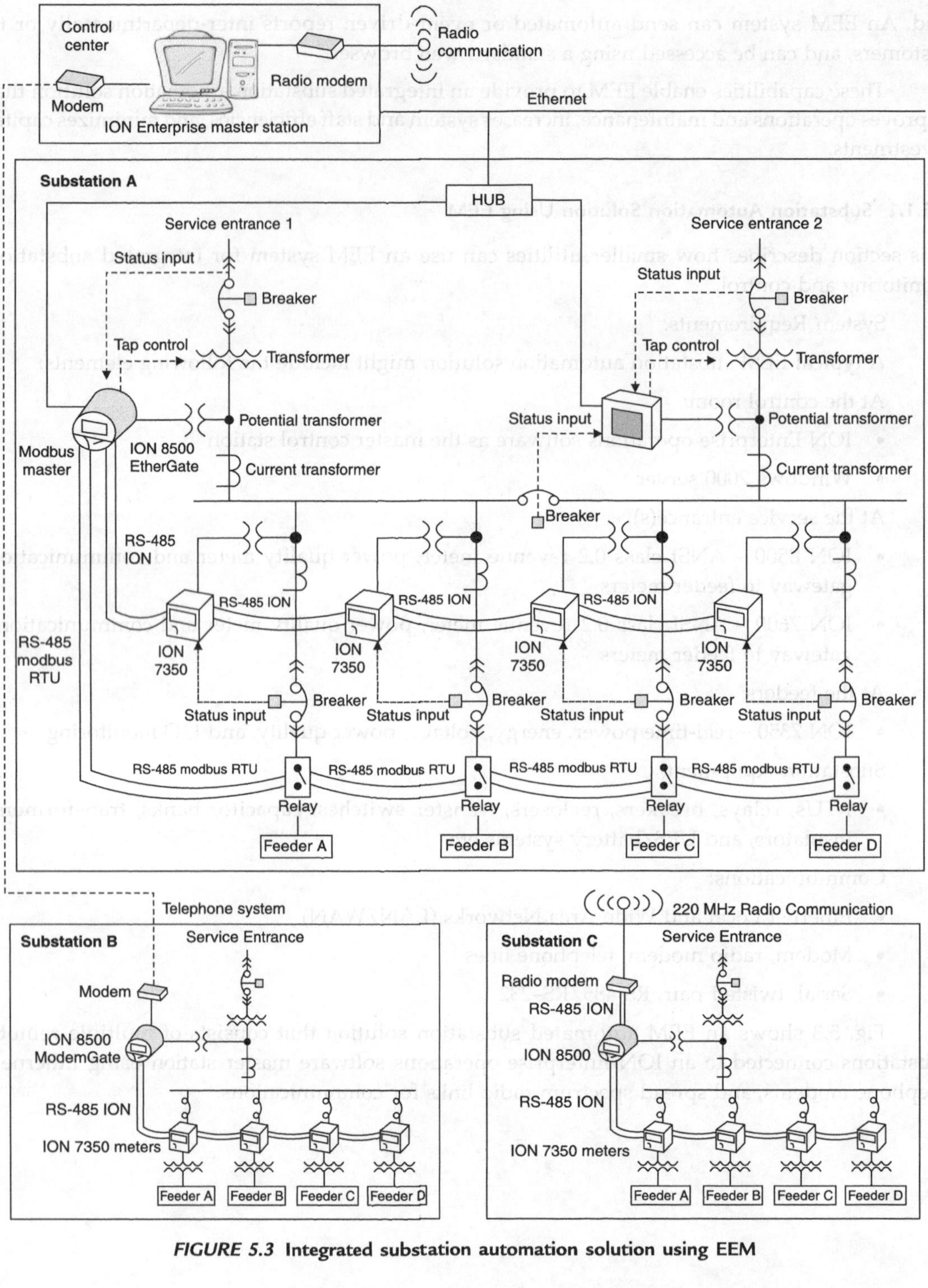

FIGURE 5.3 **Integrated substation automation solution using EEM**

5.6.1.2 An EEM System as a Counterpart to a Traditional RTU-Based SCADA System

This section describes how an EEM system integrates with a traditional RTU-based SCADA system to provide enhanced functionality.

System Requirements:

A traditional RTU-based SCADA substation automation system that incorporates an EEM system might include the following elements:

At the control room:

- SCADA master station
- ION Enterprise operations software for power quality data analysis and reporting
- Windows 2000 server

At the service entrance(s):

- ION 8500 – ANSI class 0.2 revenue meter, power quality meter, and communication gateway to feeder meters
- ION 7600 – ANSI class 0.2 revenue meter, power quality meter, and communications gateway to feeder meters

At the feeders:

- ION 7350 – real-time power, energy, voltage, power quality, and I/O monitoring

Substation equipment:

- RTUs, relays, breakers, transducers, reclosers, transfer switches, capacitor banks, transformers, insulators, and UPS/battery systems etc.

Communications:

- Ethernet Local and Wide Area Networks (LAN/WAN)
- Modem, radio modem, telephone lines
- Serial, twisted pair, RS–485/RS–232

Fig. 5.4 shows how an EEM system can be integrated with a traditional RTU-based SCADA system to share communications links and metered data, reduce substation equipment, and provide system intelligence.

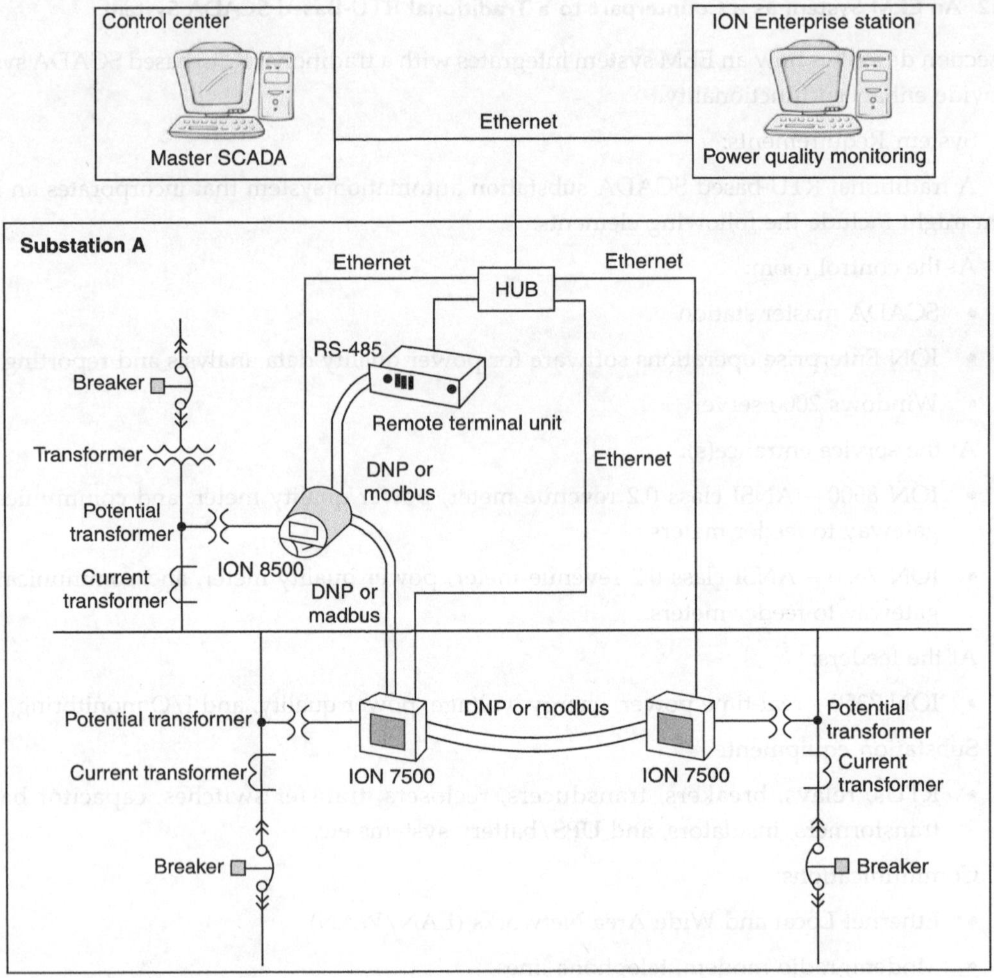

FIGURE 5.4 **An EEM system as a counterpart to a traditional RTU-based SCADA system**

5.7 ADVANTAGES OF AN EEM (ENTERPRISE ENERGY MANAGEMENT) SUBSTATION AUTOMATION SOLUTION

Utilities must manage the overall generation, transmission and distribution of electric energy in the most efficient and cost-effective manner. This entails operating, controlling, and maintaining:

- Generation assets— include controllers, relays, meters, fuel systems, cooling systems, prime mover systems, and telecom systems.
- Transmission and distribution substations—include RTUs, relays/fuses, meters, breakers, reclosers, switches, capacitors, transformers, insulators, UPS/battery systems, yard perimeter systems, and telecom systems.

- Transmission and distribution lines—include relays/fuses, over-head switches, insulators, poles, underground equipment, yard perimeter systems, and telecom systems.

An EEM system is composed of intelligent devices, communication links, and software that combine to form a layer of "energy information technology" that rides on top of key power system components and critical loads. An EEM system monitors the key electrical and physical characteristics of each power component and critical circuit within a single substation or across multiple substations, and provides operations personnel with the information they need to extract the highest possible efficiency and reliability from the system as a whole. Fig. 5.5 shows an EEM system.

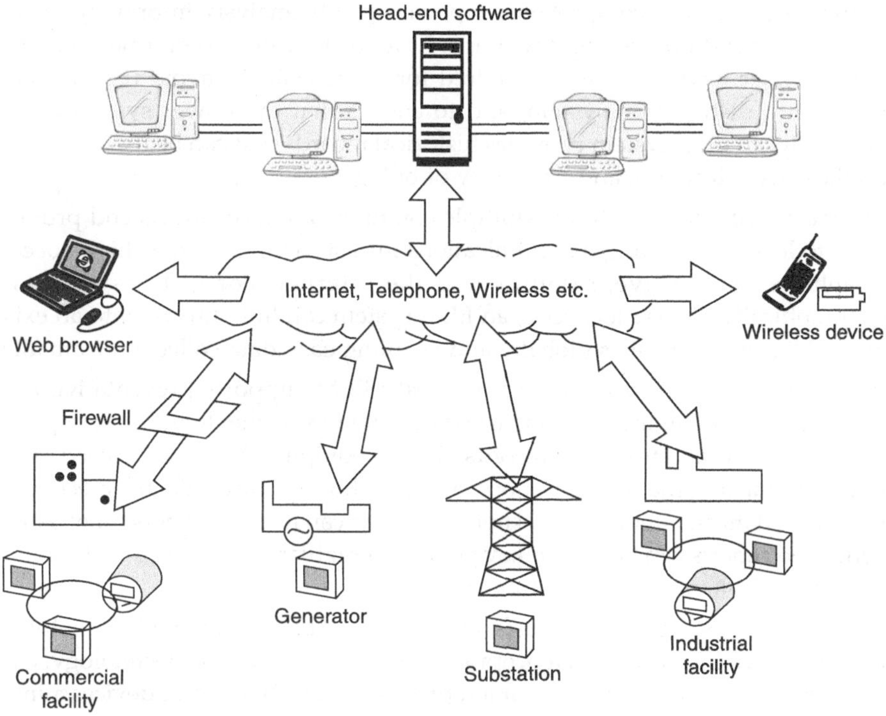

FIGURE 5.5 **Enterprise energy management system**

5.7.1 Lower Total Cost of Ownership and Added Value

An EEM system can reduce the total cost of ownership and give more value to customer base through:

- *Lower substation costs:* Intelligent ION metering and control devices reduce the number of substation components required for efficient operations. An ION meter packs monitoring, analysis, and control requirements into a single device, providing an all-in-one solution for load profiling, load aggregation, 0.2 class revenue metering, and power quality monitoring of critical circuits. ION meters provide alternative communications channels into the substation without requiring modifications to the existing RTUs or communications channels, reducing costs and risk.

- *Reduced downtime:* Critical loads and equipment are monitored "24 X 7," and the system can alarm on any combination of conditions at multiple meters or sites, enabling fast fault correction, load shedding, or generator control. Time-stamped alarms can be sent to any networked computer, email account, pager, fax or cell phone, and are automatically logged in the EEM database for sequence-of-events analysis. One ION meter can forward the alarms from many devices, and different user groups can receive event-specific notifications. EEM correlates real-time data from time-synchronized intelligent meters and other devices so impending faults can be predicted, prevented, and analyzed. EEM pinpoints faults with comprehensive power quality analysis, improving responsiveness to problems and preventing downtime. Diagnostics can be performed remotely using a web browser, saving the time required for a site visit. To avoid repeat occurrences of faults on a distribution system, and to ensure that the power system operates more efficiently, an EEM system provides historical records that can be used to analyze details on the cause, duration, and frequency of outages.
- *Lower data acquisition expenses:* Multiple communications channels and protocols enable EEM to leverage existing communications infrastructures such as high-speed Ethernet, telephone, microwave, satellite, and the Internet. For utilities with facilities in geographically diverse locations, an EEM system can link directly into an existing LAN or WAN, connecting sites globally and enabling easy data collection from all sites.
- *Lower long-term equipment maintenance costs:* EEM supports preventative maintenance and operations planning, minimizing equipment replacements and lowering maintenance costs. Real-time and historical records identify equipment that will exceed safe operating limits by tracking the number of operations or hours of use. When a piece of equipment has reached its maintenance threshold, an alarm can notify key personnel. Automatically generated reports can be sent to key personnel regularly, to assist in determining system stresses and maintenance requirements.
- *Reduced overall SCADA system costs:* PC-based SCADA used in an EEM solution is much less expensive than traditional proprietary RTU-based SCADA technology. Off-the-shelf operating systems and intelligent microprocessor-based electronic devices minimize costs, including expenses for training and system maintenance.
- *Low initial investment that can meet future needs:* With modular, industry-standard components that can be customized for specific applications, an ION energy management system adapts quickly and affordably to changing needs as facilities are added or upgraded, and power quality and operational requirements evolve. System flexibility and compatibility allow a utility to extend its ION system at its own pace, adding newer components as they become available, without interrupting or impacting existing functions.

Fig. 5.6 shows the typical benefits of substation automation in terms of % reduction of various parameters.

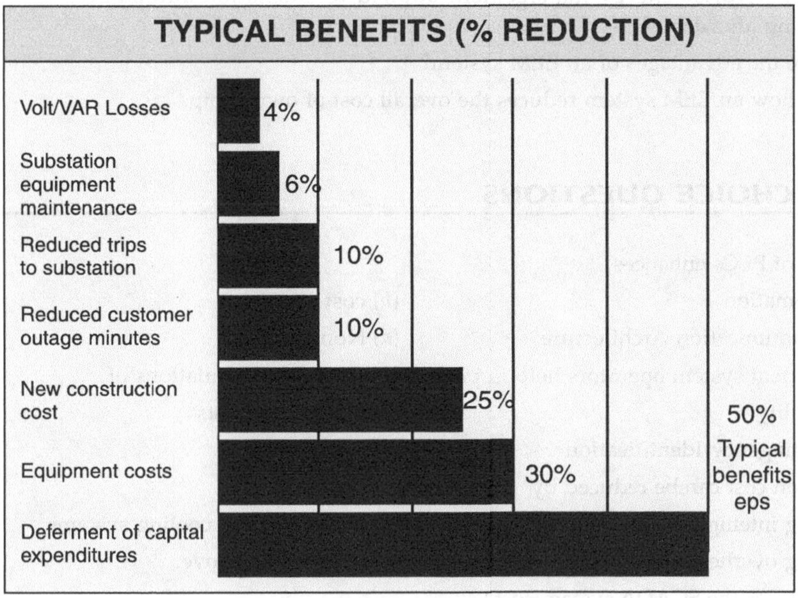

FIGURE 5.6 **Typical benefits (% reduction)**

5.8 CONCLUSION

Substation Automation has become a necessary part of the total utility automation package—particularly for utilities wanting to improve their commercial position. Substation Automation ensures power system stability, fast load restoration, advanced fault notification and fault analysis. In addition, condition monitoring can be implemented, reducing maintenance costs while improving reliability.

LONG TYPE QUESTIONS

1. Define substation automation.
2. Explain the new technologies in substation automation.
3. Enumerate the objectives of substation automation.
4. What are the functions of substation automation?
5. Explain the benefits of substation automation.
6. What are the components of a substation automation system?

7. Explain how Ethernet with TCP/ IP is incorporated in a substation automation system.
8. How can affordable substation monitoring be acquired?
9. How is enterprise energy management applied to substation automation to make substation monitoring affordable?
10. What are the advantages of an EEM system?
11. Explain how an EEM system reduces the overall cost of ownership.

MULTIPLE CHOICE QUESTIONS

1. The use of PLCs enhances
   (*a*) automation (*b*) cost
   (*c*) Communication Architecture (*d*) None of these.
2. Independent system operators help in performing real time calculations of
   (*a*) Stability (*b*) Operating Costs
   (*c*) Contingency Identification (*d*) All of these.
3. Substation cost can be reduced by
   (*a*) using intelligent ION metering (*b*) using efficient cooling systems
   (*c*) using overhead switches (*d*) none of the above.
4. EEM reduces the SCADA system cost
   (*a*) by employing PC based SCADA system (*b*) by employing proprietary RTU based SCADA
   (*c*) making use of LAN (*d*) none of the above.
5. EEM should be employed because
   (*a*) Substation costs are reduced (*b*) Data acquisition expenses are decreased
   (*c*) Long-term equipment maintenance cost is lowered
   (*d*) All of the above.

REFERENCES

1. *"Electric power distribution"* by A.S.Pabla.
2. William J. Ackerman *"Substation automation & EMS"* IEEE 1999,pages 274–279
3. William R. Cassel *"Distribution Management Systems: Functions and payback"*IEEE Transaction of Power System Vol.8 No.-3 Aug1993 pages–796–801
4. Bricker S., Gonen T. *"Substation automation technologies and advantages"* Computer Application in Power System IEEE Vol. 14 ,Issue 3 ,Jul 2001 pages 31–37
5. H. Lee Smith *"Substation Automation Problems & Possibilities"* Computer Application in Power Systems IEEE, pages 33-36.

6. Koustuv Ghoshal *"Distribution Automation: SCADA Integration is Key"* Computer Application in Power System IEEE, Jan 1997 pages 31–35.
7. Joachim Bertsch, Daniel Karlsson, *"Wide Area Protection and Power system Utilization"* Proceedings of the IEEE Vol. 93, No.-5 May 2005, pages 997-1003
8. www.vde.com
9. www.pwrm.com
10. www.geindustrial.com

Chapter 6

FEEDER AUTOMATION

6.1 LOSSES IN DISTRIBUTION SYSTEMS

6.1.1 Technical Loss

Every element in a power System (a line or a transformer etc.) offers resistance to power flow and thus consumes some energy while performing the duty expected of it. The cumulative energy consumed by all these elements is classified as **"Technical Loss."**

6.1.2 Commercial Loss

Losses occur on account of non-performing and under performing meters, wrong applications of multiplying factors, defects in CT & PT circuitry, meters not read, pilferage by manipulating or by passing of meters, theft by direct tapping etc. These are all due to non-metering of actual consumption and are called commercial losses. The total of "Technical" and "Commercial" losses is termed as **T & D loss.** It is unfortunate that in addition to the above, there is also a loss in revenue due to non-realisation of billed demand. This is in addition to commercial losses and the aggregate of T & D loss and revenue loss due to non-realisation is termed as **"AT & C loss"** (Aggregate technical and Commercial loss).

Therefore, AT & C loss to the utility is the sum total of technical loss, commercial losses and shortage due to non-realisation of total billed demand.

Example:

Units Input	100 MU	Revenue collection with reference Billed demand	90%
Unit sold	70 MU	This means out of 70 MU sales	63%

		Realization is for 90% of 70 MU	
Losses	30 MU	ATC losses	37%
% T & D loss	30%		

Power distribution systems in developing countries had to face phenomenal and rapid growth of load in the last two decades. Distribution systems were expanded on adhoc and haphazard basis keeping minimum investments in view.

Power systems are highly cost intensive and the investments needed to reduce technical losses by every 1% are too high. Computer aided load flow studies are to be made to arrive at peak power loss and also to arrive at technical losses based on load-duration curves. The Transmission system planning is done on a separate footing that is keeping in view system adequacy, system security, reliability etc for n–1 and n–2. Losses in transmission systems are too small and loss reduction is not issue of focus in Transmission system planning.

6.2 SYSTEM LOSSES AND LOSS REDUCTION

The T and D losses of the utility can be computed as

T and D Loss = (Generation + Power purchase) – (Metered + Un-metered Sale)

All the LT/HT agricultural customers, Public Water works for Gram Panchayats and C class Municipal Councils; Power loom customers have not been provided metered supply in M.S.E.B. area. The un-metered sale is a guesstimate. Basically, the T and D losses can be classified as Technical Losses and Commercial Losses. The technical losses directly depend on the line current and resistance of the conductor. The resistance is directly proportional to the length of the line and is inversely proportional to the area of cross section of the conductor.

6.2.1 Need to Curb the Commercial Losses

As per CEA's guidelines the break up of 15.5% standard T and D losses for Indian systems is as follows:

1. Gen Step up transformers EHV system, step down transformers up to 33KV system: – 4%
2. Sub transmission system step down transformation up to LT (0.4 KV system):–4.5% losses
3. LT system and service lines: – 7.0% losses

It's alarming to note that the losses in Indian system are ranging from 32–55%. The National loss of all India system is over 35%. Comparatively the T and D losses of neighboring countries are much less. (*e.g.*, Pakistan: 16.82%, China: 5.32%, Korea: 5.87%, Thailand: 6.9%).

The question arises :

Where does the commercial loss take place?

- Not in Power Stations/ EHV transmission lines
- Not in EHV/HV Substation
- Not in ICT's/power transformers

- The commercial losses take place only at consumer premises!! The land, building, installation where these losses are taking place, are belonging to the customers.

6.2.2 Importance of Investigation

The distribution systems, presently are characterized by increased power losses, inadequate metering, uncoordinated control of equipments and frequent failure of major equipments mainly due to non availability of their loading information and other status. Rapid growth in industrialization and irrigation needs have resulted in an enormous increase in the demand of electricity. All the three constituents of power system viz. generation, transmission and distribution have increased in size and complexity. Generation and transmission have expanded in a relatively planned manner using modern technology. Expansion of distribution system, on the other hand, has been mostly in an unplanned way. This has resulted in non-optimal, inadequate conductor size, highly stressed operation of equipments and lengthy, low voltage feeders. All these factors lead to following problems:

1. Increased System Losses

Results of various surveys indicate that the national average transmission and distribution energy loss is 19 %, of which the share of distribution system is 15 %. In some of the distribution networks the energy losses are 40 % or even higher. The target value of distribution energy losses is about 4 %. Hence if the distribution losses are alone brought closed to the target value, the currently reported average deficit of 8.5 % in our country between energy supply and demand can be met to a great extent without immediate addition of new generating capacity.

2. Poor Quality

Equipment failure is quite frequent in distribution system, mainly because of virtually unmonitored distribution networks and poor maintenance of distribution equipments. The operator at a substation has almost no information about the loading and voltage conditions of various feeder sections except the main feeder from the substation. The monitoring and maintenance of the distribution equipment such as transformers, lines etc., are difficult as they are remotely and sparsely located. The transformer failure rate in India is reported to be 15% as against 1 % in developed countries. Lengthy feeders have resulted in poor voltage at the tail end, thus stressing the electrical equipment adversely.

3. Increased Peak Demand and Unbalanced Loads

The existing distribution system has limited control over its feeders. Hence proper scheduling of loads cannot be undertaken. This results in steep rise in loads during peak hours and unbalanced operation for most of the time.

4. Poor Return of Revenue

Present practices of manual meter reading, delay in billing, faulty and inaccurate metering, tampering of meters, and pilferage of electricity are some of the main reasons for poor return of revenue to electricity boards.

5. Breakdowns

Distribution feeders may also remain out of service due to breakdowns. This break downmay be because of faults and some abnormal behavior of system. It results into complete feeder outage losing considerable revenue.

6. Overloading of Feeders

Normally new loads on feeder are added without thinking of it's over stretching. This results into higher system losses, over heating of conductors, poor voltage profile and ultimately over loading of transformer. At the same time, some of the feeders are under loaded. Hence the reconfiguration is necessary under such circumstances.

6.2.3 Steps Taken To Reduce Line Losses

(*a*) **Energy accounting.** The energy accounting activity has been given high priority since April 2000. The initiatives taken so far are as follows:

- *Energy accounting of industrial areas:* The energy accounting for industrial areas have been started from January 2001 and feeder wise energy balance from various substations is being undertaken for evaluation of losses. 78 MIDC areas have been identified for loss monitoring.
- *Energy accounting of express feeders:* Energy accounting is carried out for express feeders exclusively feeding HT consumers. Attention is given to minimize losses on these feeders.
- *Energy loss evaluation at EHV level:* State level energy balance is established to find the energy sent for distribution from EHV system *i.e.*, 400V to 66kV.
- *T and D loss evaluation at Division level:* Division level energy accounting has been started and evaluation of losses at all the distribution divisions is carried out on a monthly basis.

(*b*) **Metering**. There has been a quantum jump in the amount to be spent on procurement of meters. However, the success of implementation of the ambitious metering program would require an environment favoring such conversion. This necessitates upfront correction in tariffs of Agriculture, Power loom, and PWW categories. The tariffs should be designed to incentives consumers to shift from flat rate tariffs to metered tariffs. Any premature effort to forcibly convert flat rate tariff consumers to metered categories is bound to recoil and be counter-productive to the objective of the metering program.

(*c*) **Daily disconnection reports**. From 15th January 2001, the Board has prescribed a daily disconnection report from the field staff. As a result, the number of disconnections has gone up substantially. Daily visits have improved detection of thefts also.

(*d*) **Detection of thefts.** In spite of best efforts, the incidence of use of electricity without the intention of paying for the same is on the increase, resulting in loss of several crores of Rupees to the Board. The present criminal and civil legal provisions are not sufficiently deterrent to defaulters. In order to curb this menace, the Board has already suggested to the Government,

1. To amend the Indian Electricity Act, 1910 for application in the State of Maharashtra.
2. To enact a bill titled "Maharashtra State Electricity Theft & Malpractices (Prohibition) Bill" for constitution of Special Tribunals and Special Court to try cases of criminal and civil mal-practices. The amendments to the Indian Electricity Act, 1910 suggested are;
   - To search and seize or confiscate any equipment used for such unlawful activity by empowering the authorized officer of the Board with necessary powers.
   - Criminal cases to be tried summarily by inserting a suitable provision after Section 49 A under the heading 'Criminal offences and proceedings' in the Indian Electricity Act, 1910. In case of theft, under current legal provisions, both criminal and civil proceedings are required which are dealt with by different agencies. As theft of electricity has to be dealt with by different agencies, there are extended delays associated with the same. It is suggested to enact a separate law to constitute Special courts with necessary powers to award compensation to the Board and take necessary criminal action against defaulters.

(*e*) **Tariff correction**. The present tariff structure subsidizes 9 out of 10 consumers. The Board has 1.29 Crore consumers, of whom 1.16 crore receive energy at less than cost of supply. The result is that the 10th consumer has to pay an abnormally high tariff. This makes theft attractive and collusion profitable. Collusive theft is difficult to detect. On the other hand, abnormally low tariff encourages waste. Physical policing of 9 crore population and 6 Lakh km of lines has its limitations. The answer lies in correcting tariff. Tariffs also need to be corrected to incentives shift of flat rate tariff consumers to metered categories. Further, the tariffs should have a large proportion of fixed charges consistent with the Boards fixed expenses. Tariff structure should also aim to utilize Demand Side Management (DSM) initiatives with an aim to flatten the load curve.

6.3 NETWORK RECONFIGURATION

6.3.1 Introduction

Radial networks have some advantages over meshed networks such as lower short circuit currents and simpler switching and protecting equipment. On the other hand, the radial structure provides lower overall reliability. Therefore, to use the benefits of the radial structure, and at the same time to overcome the difficulties, distribution systems are planned and built as weakly meshed networks, but operated as radial networks. The radial structure of distribution networks is achieved by placing a number of sectionalizing switches in the network (usually referred to as tie switches) used to open the loops that would otherwise exist. These switches, together with the circuit breakers at the beginning of each feeder, are used for reconfiguration of the network when needed. Obviously, the greater the numbers of switches, the greater are the possibilities for reconfiguration and the better are the effects. Generally, network reconfiguration is needed to: (*i*) provide service to as many customers as possible following a fault condition, or during planned outages for maintenance purposes, (*ii*) reduce system losses, and balance the loads to avoid overload of network elements.

Reconfiguration means, altering the topological structure of distribution feeders by changing the open/close status of sectionalizing switches in the feeders , under both normal and peak load conditions.

6.3.2 Benefits of Reconfiguration

- Restoring power to outage portion of a feeder in a timely manner which improves the value of service to customer by reducing average outage.
- Relieving of overloads on feeders by shifting load in real-time to adjacent feeders which could defer capital expansion projects.
- Reducing resistive line losses, which could reduce the operating cost of a distribution system.

6.3.3 Approach for Reconfiguration

There have been a number of works concerning resistive line losses reduction in distribution networks through reconfiguration. Generally there are two approaches to the reconfiguration problem.

The first approach would be to determine the status of all switches in the network simultaneously. Due to combinatorial nature of the problem, very complicated mathematical techniques should be used and large computational time is needed. The second approach would be to deal with each possible loop (determined by an opened tie switch) one at a time.

Methods based on this approach are simpler and faster.

Traditional optimal configurations are obtained by minimizing power losses. For a given period, a moment of time is taken as a standard (usually the system peak) and power loss optimization method is used to determine the configuration of the network. The problem of minimization becomes very complex if energy losses are to be minimized. Since loads change on an hourly basis or even shorter, configuration of the network may need to be changed accordingly.

6.3.3.1 Based on Either of these Approaches Following are theTechniques of Reconfigurations

(A) Power loss Minimization Techniques:

- State of the art techniques:
  - (*i*) Optimization techniques
  - (*ii*) Heuristics based methods
- Artificial intelligence (AI) Based methods

(B) Energy loss Reduction technique.

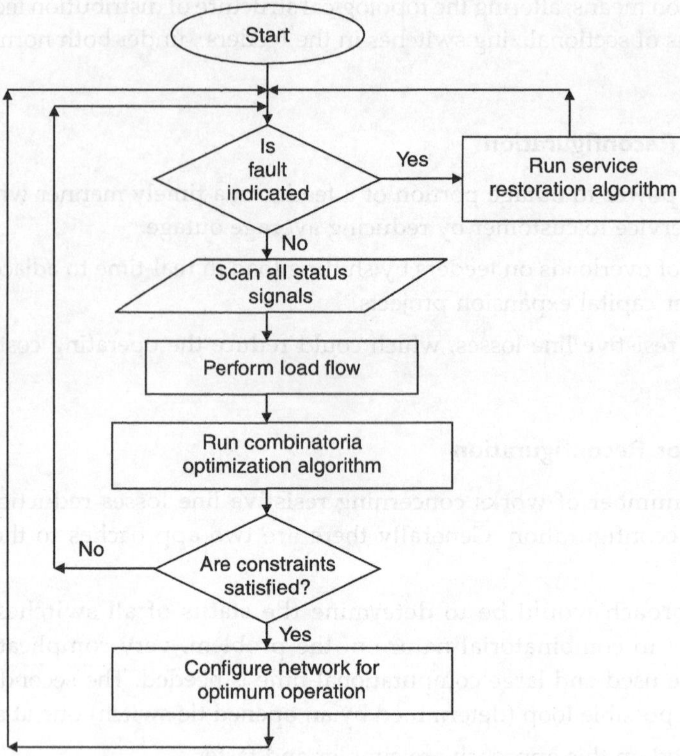

FIGURE 6.1 **Network reconfiguration algorithm**

6.4 IMPROVEMENT IN VOLTAGE PROFILE

Network reconfiguration is performed by altering the topological structure of distribution feeders. By reconfiguring the network, voltage stability can be maximized for a particular set of loads in distribution systems. A new algorithm is formulated for enhancement of voltage stability by network reconfiguration. Initially, a certain number of switching combinations are generated using the combination of tie and its two neighbouring switches, and the best combination of switches for maximizing the voltage stability in the network among them is determined. Then considering the branches next to the open-branches of the best configuration one by one to check whether there is any other switching combination available, which gives further improvement in voltage stability, extends the search. The enhancement of voltage stability can be achieved by the proposed method without any additional cost involved for installation of capacitors, tap-changing transformers and the related switching equipment in the distribution systems. The method has been tested on a 69-bus test system, and the results indicate that it is able to determine the appropriate switching options of the optimal (or near optimal) configuration for voltage stability enhancement with less computation. It has also been shown that power losses are reduced when voltage stability is improved by network reconfiguration.

Recently, there has been a growing interest in optimizing the operation of distribution networks particularly in the area of distribution system automation. Distribution systems are normally configured radially. From time to time, modifying the radial structure of the feeders by changing the *on/off* status of the sectionalizing and tie switches to transfer loads from one feeder to another may significantly improve the operating conditions of the overall system. As the distribution systems normally have a combination of industrial, commercial, residential and lighting loads and the peak load on the substation transformers and feeders occur at different times of the day, the systems become heavily loaded at certain times of the day and lightly loaded at other times. If the distribution loads are rescheduled more efficiently by network reconfiguration, the voltage stability in the system can be improved. Reconfiguration also allows smoothening out the peak demands, improving the voltage profile in the feeders and increasing the network reliability. The voltage stability phenomenon has been well recognized in distribution systems. Several analytical techniques have been proposed to assess the risk of voltage instability. Voltage stability is important for power system operation and planning because voltage instability may lead to voltage collapse and hence outage and monetary losses, and possibly total system black out. Furthermore, a stable system contributes to reliability and reduction in system loss. Voltage instability occurs when the power system is subjected to heavy loading of transmission lines and insufficient local reactive power supply. A relationship between voltage stability and power losses has been proposed to show that voltage stability is improved when power losses are reduced. Here, a new scheme for the automation of distribution networks for voltage stability enhancement by network reconfiguration is studied. The best configuration for maximum improvement in voltage stability in the system is obtained by considering all possible branches which are candidates for branch exchanges. The enhancement of voltage stability is obtained by the given method by reconfiguring the network structure without any additional capacitor installation, tap-changing transformer and related switching equipment.

6.4.1 Mathematical Formulation of Voltage Stability Measurement

By applying the single-line method for the reduction of distribution network, the occurrence of voltage collapse can be studied easily, and it is not necessary to consider every line of the network separately. For a reduced single-line network, the voltage stability index has been derived in and is reproduced as,

$$L = 4\left[(x_{eq}p_{leq} - r_{eq}Q_{leq})^2 + x_{eq}Q_{leq} + r_{eq}P_{leq}\right] \tag{1}$$

where, P_{leq}, and Q_{leq} are the total real and reactive loads respectively. The equivalent resistance and reactance of a reduced single line network have been defined and rewritten,

$$r_{eq} = \Sigma P_{loss}/\{(P_{leq} + \Sigma P_{loss})^2 + (Q_{leq} + \Sigma Q_{loss})^2\} \tag{2}$$

$$x_{eq} = \Sigma Q_{loss}/\{(P_{leq} + \Sigma P_{loss})^2 + (Q_{leq} + \Sigma Q_{loss})^2\} \tag{3}$$

It is noted that for a stable system, the value of stability index, L is very much less than 1.0. However if the value of L approaches 1.0, this would indicate that the system is close to voltage collapse. If the network is loaded beyond this critical limit, the power becomes imaginary and it is at this point that the voltage collapse occurs.

6.4.2 Enhancement of Voltage Stability by Network Reconfiguration

In a radial distribution system, network reconfiguration is performed by closing / opening the tie and sectionalizing switches. However, it may lead to infeasible configuration by forming a closed loop or leaving out one or more branches unconnected. This can be avoided by checking the connectivity from the source to all the nodes during the network reconfiguration.

(A) Determination of Switching Combinations

Fig. 6.2 shows a tie branch with its two neighbouring branches forming a switch group. The number of switch groups will be equal to the number of tie branches in the system. When a sectionalizing switch is opened for reconfiguration, it will play the role of a tie switch in the next stage where this open branch and its adjoining branches again form a switch group.

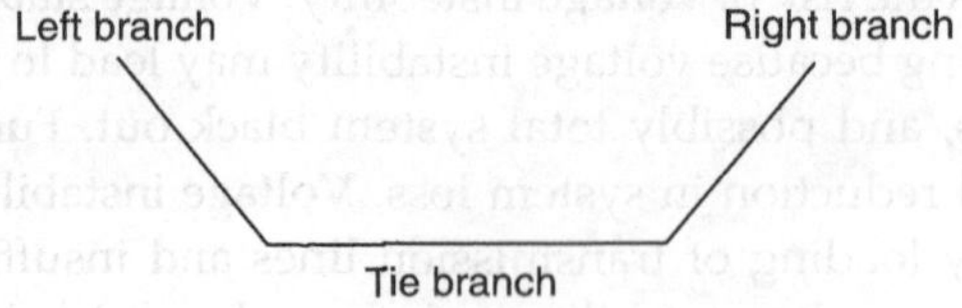

Three branches forming a switch group

FIGURE 6.2 **Branches forming a switching group**

The above recursive procedure is terminated when a node with three or more branches is encountered, since the other branches will be considered in the formation of the switch group of the respective feeders. This constraint is imposed to avoid the combinatorial explosive nature of the problem.

The switching options in a switch group can form three combinations with one of them only open for each combination. The three combinations can be named as zero connect (0), left-connect (1) and right-connect (2), where, zero-, left- and right-connect represent the configuration at which the middle, right and left branches are open respectively. 0, 1 and 2 are used to represent connectivity of the switches during formulation of the switching combinations.

Let us consider all the switch groups in a system. If there are *n* tie switches in the system, there will be *n* switch group forming 3" different switching combinations. Each switch position in a switching combination is found out by using the formula,

$$SW_i = [P_{ij} \ldots, P_{i2} + P_{i1}] \tag{4}$$

And the switch positions of the jth switch in ith combination, ***Pij,*** can be obtained as,

$$P_{ij} = [(i-1)/3^{(j-1)}] \bmod 3 \tag{5}$$

where, *i* is the combination number (one of 3"), *j* is the switch number (one of *n*), and *SW*, is the positions of the various switches in the ith switching combination. The status of each switch in a switch group is realized for each switching combination. For zero-connect, the configuration will remain unchanged, but for left-connect or right-connect, the configuration will change according to the left or right-connect logic. Left and right-connect logic are developed using the power flow in the adjoining branches of the tie branch in the system. When a system is reconfigured by closing the tie branch and opening either of the adjoining branches, the power flow in the open-branch, which was flowing before reconfiguration, shifts to the newly connected tie branch. If the left-connect is performed, then the power flow of the right branch of the tie branch will be shifted to the tie branch. Similarly, when the right-connect is performed, the power flow of the left branch of the tie branch will be shifted to the tie branch.

(B) Permissible Switching Combinations

A distribution system after reconfiguration may become infeasible because there can be some loads which may be isolated from the source of power and some closed loop may be formed. The infeasible switching combinations are determined based on maintaining radiality and avoiding isolated lines. This can also be extended to include operationally imposed switching constraints such as plant/cable ratings, temporary embargoes, etc. These infeasible combinations must be removed from the total switching combinations to obtain the permissible switching combinations.

The infeasible combinations are evaluated by checking the connectivity of each branch in the network using the connectivity logic explained below: (*i*) Check the existence of physical connections of a node or branch from the source. If it is feasible, assign the connectivity to be active. (*ii*) Repeat the above step for all nodes in the network. (*iii*) Repeat steps (*i*) and (*ii*) for all the switching combinations obtained using eqn. 4. The above steps can be represented by the following recursive procedure:

$i \leftarrow 1$

CC(i)

{ $con[i] \leftarrow 1$

$TC[i]$: CC(for[i]); }

If $P_{ij} > 0$ *then for* $[i] = j; rev[j] = i;$

If $P_{ij} < 0$ *then for* $[j] = i; rev[i] = j;$

$i \leftarrow 1$

CC(i)

{ $con[i] \leftarrow 1$

$TC[i]$: *CC(for[i])*; }

If $P_{ij} > 0$ *then for* $[i] = j; rev[j] = i;$

If $P_{ij} < 0$ *then for* $[j] = i; rev[i] = j;$

where, CC(i) is the connectivity check function, *con[i]* is the connectivity of node i, *TC[i]* is the total numbers of connection for ith node, P_{ij} is the power flow from sending node i to the receiving node j , and *for* and *rev* are the forward and reverse nodes respectively.

To find all the connections of the node i, the following procedure is used:

If for[i] = n: and rev[j] = y, then con[i] = x and y; ***or*** *x and a; or i and y.*

The connectivity check for the whole system is done as follows:

Con[i] = 1; *Pi* V Qi # *NULL*

Infeasible con[i] = 0; ***i*** = 1,. . . , n

Check whether the connectivity flags of all the nodes of that particular configuration are active; otherwise declare the combination as infeasible.

Vi con[i] # 0, Feasible

(C) Automating the Search Process for Voltage Stability Enhancement

For the process of automation of the search, a method has been developed to identify the best combination among all the permissible switching combinations. This solution may not be the optimal or near optimal since the search has been restricted to only a limited number of combinations. In the next stage of automation, the search is extended to find the global optimum for voltage stability enhancement by moving to the left or right of each of the open or tie branches, depending upon whether the right or left adjoining branch of the tie branch is open. This is done by considering the open sectionalizing or tie branches as the tie branches for the next operation of the reconfiguration.

From the switching status of each switch group obtained in the best combination, the connectivity of the switch group is examined for the identification of zero-, left-, or right-connect. The switch group remains unaltered for zero-connect. The connection is moved towards right (if it is left-connect) or left (if it is right-connect), by opening the next neighbouring branch of the left or right of the current tie branch respectively, and closing the current tie branch to form a new switch group. The stability value for the newly formed combination is compared with that of the best combination found earlier, and if it is less, the current combination is considered as the best combination and the search is further carried out in the same direction; otherwise the best combination remains the same. The other switch groups are also examined in succession by applying the above procedure and the overall best combination and its corresponding configuration is obtained for the maximum improvement of voltage stability.

6.4.3 Test Results and Discussions

The test system is a 69-bus hypothetical 12.66 kV radial distribution system with 7 laterals, which has been taken from a portion of the PG&E distribution system. The system consists of 5 tie lines, and there are sectionalizing switches on every branch of the system. The schematic diagram of the test system is shown in Fig. 6.3. The total system loads for the base configuration are 3802.19 kW and 2694.60kVAr. The software developed for the proposed automated scheme has been implemented using Borland C++. The program reads the data pertaining to the base configuration of the system and then generates 3^n switching combinations as shown in Table 6.1, where n is the number of tie branches. The permissible switching combinations are identified by the connectivity logic. From the simulations of the test system, it is found that there are 54 infeasible and 189 feasible combinations among the total combinations of 243. Among the permissible combination the best combination for voltage stability has been obtained for the test system. Investigation is extended to check whether any other combination gives further improvement in voltage stability. The search has been carried out by moving towards left or right of each open branch of the best configuration as the case may be. The configuration, which gives the minimum value of the stability index, is declared as the best solution for the enhancement of voltage stability.

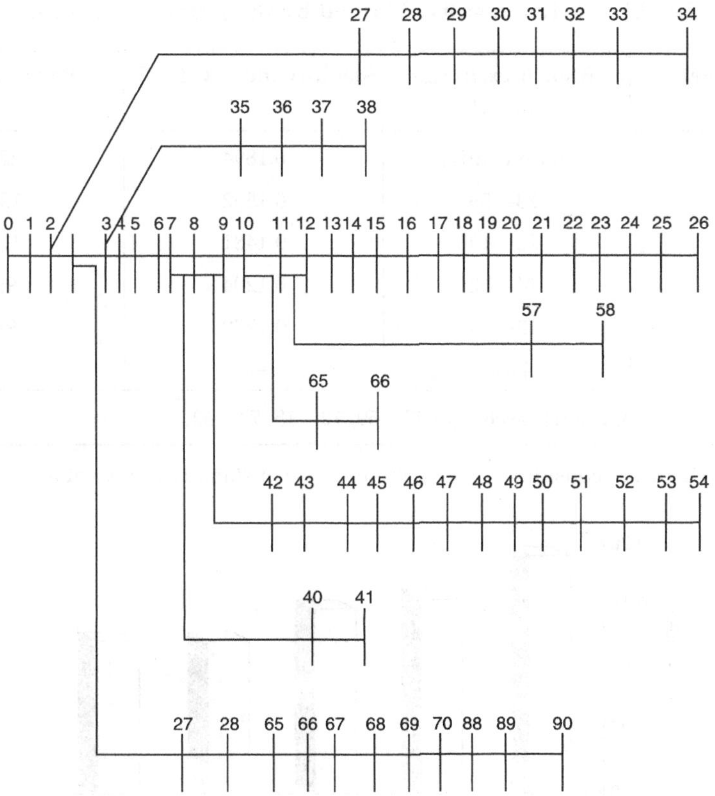

FIGURE 6.3 **Radial distribution test system with 7 laterals**

TABLE 6.1. Generation of switching combinations

Number	1	2	3	...	3^n
Switch status	00000	00001	00002	...	22222

The proposed method has found the best switching options solution for voltage stability improvement in the above test system and the results are given in Table 6.2 and Fig. 6.4. Test results show that the proposed algorithm can give the appropriate switching options for voltage stability enhancement within a reasonable computer time and effort for network of any size and complexity. At every search level of the proposed algorithm, the power losses have also been investigated and it is found that when voltage stability is improved power losses in the system are reduced. Table 6.2 shows the real power loss at every search level carried out to find the voltage stability enhancement by the proposed method. It is observed that most of the bus voltages have been improved after network reconfiguration for voltage stability. The test results show that before reconfiguration, the minimum bus voltage was 0.9092 p.u. at the end of line 65, and after reconfiguration it has increased to 0.9428 p.u. at the end of line 61.

TABLE 6.2. Test results obtained by the proposed method

Search level	Branch exchange (in-out)	Stability index value L	Power loss (kW)
0	Base config.	0.1894	225.05
1	73—59	0.1552	134.07
2	72—13	0.1481	122.99
3	74—62	0.1204	99.86
4	13—15	0.1200	99.66
5	*None	—	
Selection of final branch exchange 73—59, 72—15, 74—62			

*No branch exchange in considered due to the increase of stability index value

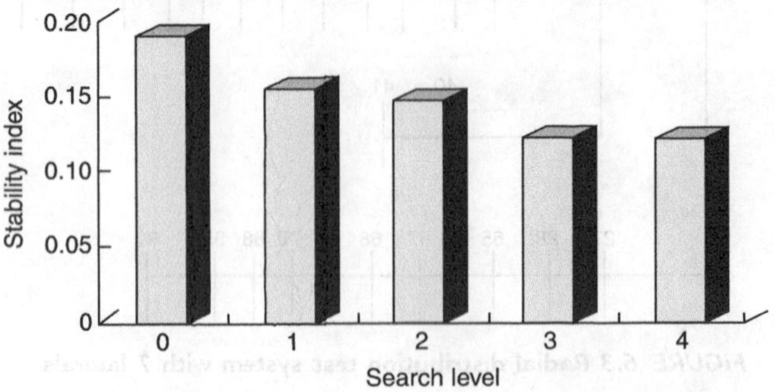

FIGURE 6.4 **Voltage stability development**

6.5 CAPACITOR PLACEMENT OF DISTRIBUTION SYSTEM FOR REACTIVE POWER COMPENSATION

6.5.1 Introduction

The reactive power is consumed by overhead lines, transformers and loads. Its proper generation and control is important for maintaining the network voltage under normal and abnormal power system operation and to reduce system losses.

Cost per KVAR of reactive power generation varies between 2 and 4% of per KW installed generation capacity and will rise with increasing use of automatic and regulating controls. The main components responsible are

Lines

(*a*) The distribution lines consume reactive power (I^2 x) depending upon the series reactance (x) and load current (I). The series reactance of line is proportional to the conductor self-inductance, which decreases as the spacing between conductors decreases.

(*b*) Increasing the line voltage at transformer and motors above the rated voltage will increase the consumption of reactive energy.

Distribution Transformers

The reactive power consumption takes place in the series leakage and the shunt magnetizing reactance. The second component is voltage independent, while the first component is proportional to the square of the transformer current. A completely unloaded transformer would be very inductive and has a very low power factor.

Loads

Loads have varying reactive power requirement. The low power factor can be a result of the equipment as in case of welders or it can result from the operating conditions under which equipment is used, as lightly loaded transformers and lightly loaded induction motors are the worst offenders.

6.5.2 Monitoring Of Reactive Power

(*a*) Reactive power monitoring is desirable to maintain efficient system operation. This is exercised by recording daily the station bus spot power factor at minimum and maximum loads.

(*b*) Another method of reactive power check is made by noting daily hours of capacitor banks utilization. Total numbers of hours during a day or a month are successively compared to check the quality of reactive compensation.

6.5.3 Compensation

Mainly capacitors are used to develop reactive power near the point of consumption for capacitor compensation at load; the user reaps the same advantage as the power utility for higher power factor on small scale. Also, if each load is compensated, the power factor remains relatively constant since in plants, loads are switched on and off and the dangers of over-compensation do not exist. If power factor has been corrected only at the service entry, system power can make relatively wide swings as heavy loads are switched on and off. Suitable capacitor banks at grid or main substation are desirable to feed reactive requirement of lines, transformers, and domestic consumers etc. who have no capacitors at terminals.

At the point of capacitors, over compensation, the voltage rises and capacitive KVAR will export *i.e.*, It will flow to reverse side of the load. Over compensation at consumer premises is not generally desirable. Minimum capacitor size is determined as

Minimum capacitors in KVAR = (Transformer KVA × % Impedance)/200

6.5.4 LT Capacitors Installation

LT capacitors are installed on distribution system on individual lines or consumer motors to reduce system losses and improve system voltage and capacity. The other advantages provided for consumer are reduction in KVA demand, losses and stable voltage LT. Capacitor on individual service is preferred because,

1. It can be controlled by same control gear which controls the motor.
2. If capacitors are connected on LT motor connection the reactive loading current flowing through NT and LT lines will improve power factor at both ends.
3. The capacitors being directly connected to motors come into operation when motors are in service and are automatically out when motors are not in use.

Capacitor placement of distribution system for reactive power compensation distribution system are critical links between the utility and customer. Hence they are designed to be most efficient at peak load demand. A typical distribution feeder involves three phase or single phase lateral lines. Under normal operating conditions, the optimum operation can be obtained when the network presents:

(*i*) Minimum voltage and frequency deviations at consumer feeding points.
(*ii*) Minimum losses.
(*iii*) Unity power factor.
(*iv*) Maximum reliability.

In an industry only the real power is useful in energy conversion and the excess wad current represents a waste to consumer, who has to pay not only for the excess cable capacity to carry it but also for the excess joule loss produced in the supply cables.

In distribution systems, as most of the loads are inductive in nature. Excess reactive power leads to:

(*i*) Lower power factor.
(*ii*) Malfunctioning of several switchgear.
(*iii*) Saturation of transformers.
(*iv*) Excessive neutral currents.
(*v*) Oscillating torque in AC machines
(*vi*) Increased ripple in rectifiers.

All this has a cumulative effect in increasing the monetary losses in the distribution system. Stable voltage profile is very important in a distribution system network but reactive power leads to loss of control of voltage in supply systems. Many drives such as pumps, compressors exhibit constant power characteristics over a voltage range of 80% to 110% of the nominal, under reduced voltage operations due to increased current requirement. Hence to maintain a stable voltage profile, reactive power compensation is required.

The means normally used for reactive power compensation and voltage control are installation of shunt capacitors, synchronous condensers, station capacitor banks and distribution capacitor banks. They reduce series reactance and shunt susceptance of line and act as line compensators. Capacitors to use in distribution system. In most cases the control of capacitor bank has been manual, which results in inefficient use of capacitor banks, when distribution stations are under-compensated during peak periods and over compensation required, it is required to find the appropriate location of capacitors, the optimum connection scheme, and the number of capacitor so that the installation is efficient as well as economical.

6.6 PROPOSED ALGORITHM FOR LOCATION OF CAPACITOR

Basically the total losses at each branch in the distribution system are calculated from load flow. The sensitivity factors of various feeder buses for a given load pattern are also calculated. The bus having the highest sensitivity is determined and the value of that point is determined. Since priority list of sensitive buses is already present. These buses serve as the candidate location initially for capacitor placement. As a result, combinatorial approach is automatically avoided. The cost of placing the capacitor at that location can also be determined.

6.6.1 Formulation of Problem

Objective

1. To improve the voltage profile by reducing the power losses and by reactive power compensation.
2. To minimize the annual cost of the system.

Assumptions Made

1. Three phase balanced network is considered.
2. Maintenance costs are not taken into account.

The voltage magnitude at each bus must be maintained with in the limits expressed as follows.

$$V_{min} <= V_i <= V_{max}$$

where

V_i is voltage magnitude of ith bus.

V_{min} and V_{max} are bus minimum and maximum voltage respectively.

Mathematically, the objective function of problem can be described as,

Minimum F = minimum (COST)

where COST includes cost of power loss and capacitor placement.

Since load flow is available, the power loss and at each line section is calculated as:

$$\mathbf{P_{loss} = [(P_i^2 + Q_i^2) / V_i^2] * R_i - 1, i}$$

where

P_{loss} is total I^2R loss in the network.

R_i–1,i is the resistance of line section between node i–1, i

P_i and Q_i are power flows through line section

V_i is the voltage at the sending end of the section

The total power loss of the feeder P_t, loss is summation of loses of all the line section of feeder.

$$\mathbf{P_t, loss = \Sigma P_{loss}(i, i + 1)}$$

The purpose of placing capacitor along distribution feeder is to lower the total power loss and bring the bus voltages within their specified limits while minimizing the total cost.

Sensitivity Analysis

The sensitivity analysis is employed to select the candidate location for placing capacitor in distribution system.

The sensitive factor associated with a bus is defined as:

$$S = P_{loss} / Qc_i$$

where

Qc_i is the reactive power at particular ith location.

S is the sensitivity factor estimated at each bus every time a configuration change occurred in feeder.

The bus having highest sensitivity factor is a ranked first in the priority of buses where the capacitor need to be placed. *i.e.*, where maximum reactive power compensation is required .The total cost and bus voltages are then calculated. If both the total cost is reduced and voltage limits are satisfied, the capacitor addition is accepted, otherwise next location is checked.

Capacitor Value

To determine the value of capacitance required for compensation of reactive power the formula used is:

$$C = (Q_s - Q_r)/V_s^2 * \omega$$

where

C is value of capacitor required

Q_s is sending end reactive power

Q_r is receiving end reactive power

V_s is sending end voltage

and $\omega = 2 \Pi f$ F- supply frequency

Cost Estimation

The cost of installing the capacitor is

$$COST = K_p P_{t'} \text{loss} + \Sigma (K_{ic} Q_{oc})$$

where K_p = Equivalent annual cost per unit of power loss

Q_{oc} = smallest size of capacitor available

K_{ic} = equivalent annual cost per KVAR

The use of capacitor will improve the power factor of the system, cost saving will be realized and distribution losses will be minimized. Capacitor compensation will reduce the time spent manually in operating banks of long feeder networks.

Advantages of Reactive Power Compensation:

- The use of capacitors improves the power factor.
- Cost saving is realized
- Distribution losses are minimized

Thus, *capacitor compensation will reduce the time spent manually in operating banks of long feeder networks. The above-proposed method can obtain accurate optimal solution with less computational time*

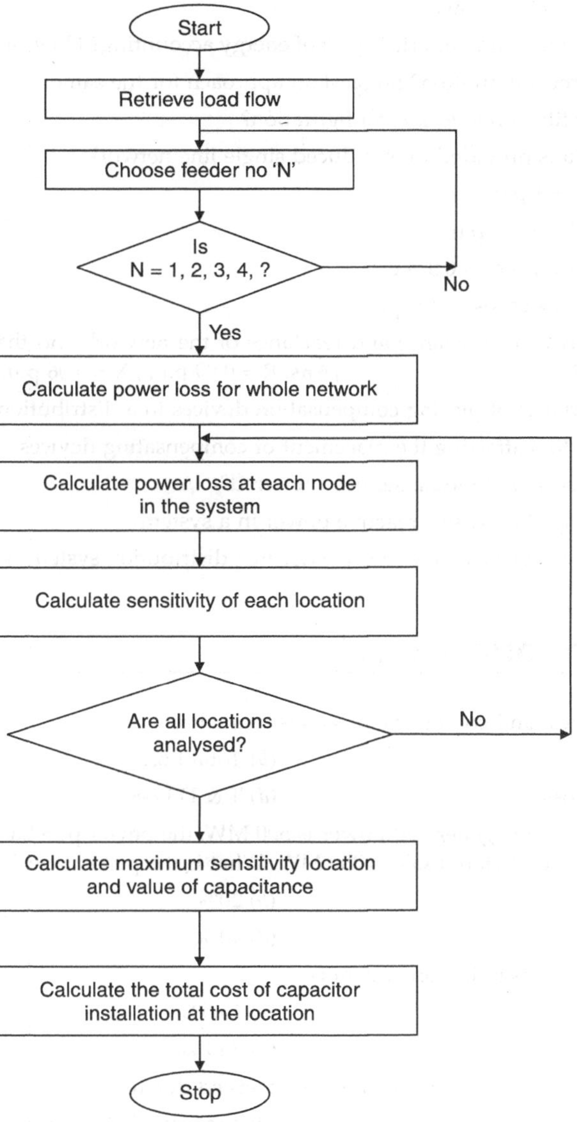

FIGURE 6.5 **Flow chart for Reactive Power Compensation**

LONG TYPE QUESTIONS

1. Define the following:
   (*i*) Technical Loss (*ii*) Commercial Loss
   (*iii*) T & D Loss (*iv*) AT & C Loss.
2. What is energy accounting?
3. What necessitates theft of power?
4. Why is detection of theft an important part of energy accounting? How can theft of power be curbed?
5. What is network reconfiguration? Suggest an approach for the same.
6. What are the benefits of network reconfiguration?
7. The following data is provided for a reduced single line network:
   (*i*) Total real load = 1 p.u.
   (*ii*) Total reactive load = 1 p.u.
   (*iii*) Total active power loss = 0.3 p.u.
   (*iv*) Total reactive power loss = 0.2 p.u.
   Calculate the equivalent resistance and reactance of the network and the voltage stability index. Is the system stable? [**Ans.** R = 0.09 p.u. ; X = 0.06 p.u. ; L = 0.6 ; system is stable]
8. Explain the importance of placing compensation devices in a distribution network.
9. Enumerate the factors affecting the placement of compensating devices.
10. Why are capacitors on individual service preferred?
11. What are the effects of excessive reactive power in a system?
12. Discuss a method for placement of a capacitor in a distribution system with a neat flow diagram.

MULTIPLE CHOICE QUESTIONS

1. The total of technical and commercial losses is termed as
   (*a*) AT & C Loss (*b*) Total Loss
   (*c*) Cumulative Loss (*d*) T & D Loss.
2. In a hypothetical system, generated power is 500 MW, the power purchased is 200 MW, the metered sale is 200 MW and unmetered sale is 250 MW. What is the percentage T & D loss?
   (*a*) 10 % (*b*) 20%
   (*c*) 30% (*d*) 40%.
3. The average T & C Loss in Indian system is
   (*a*) 35% (*b*) 15.5%
   (*c*) 16.82% (*d*) 5.87%.
4. A steep rise in the load during peak hours is observed due to
   (*a*) limited control over the system feeders (*b*) abnormal behavior of system
   (*c*) improper metering (*d*) All of the above.

5. While performing energy accounting, attention is paid to
   (*a*) increase system loss
   (*b*) decrease system loss
   (*c*) ensuring proper metering
   (*d*) reducing breakdowns.
6. Major portion of T & D losses is contributed by
   (*a*) line loss
   (*b*) losses in the substation
   (*c*) theft
   (*d*) all of the above.
7. Network reconfiguration is achieved by
   (*a*) placing transformers on a system line
   (*b*) placing circuit breakers.
   (*c*) placing proper number of sectionalizing switches
   (*d*) none of the above.
8. The equivalent inductance and resistance of a reduced single line network is 0.5 and 0.5 p.u. The total real and reactive loads are both 1 p.u. Is the network stable?
   (*a*) Yes
   (*b*) No
   (*c*) Cannot be determined.
9. The load current in a distribution line is 20 A when the series reactance is 50 ohm. The reactive power consumed is:
   (*a*) 10 kW
   (*b*) 20 kW
   (*c*) 30 kW
   (*d*) 40 kW.
10. Two distribution lines have series inductances of 50 and 55 ohm for load currents of 20 A and 25 A respectively. The ratio of reactive power consumed is:
    (*a*) 0.85
    (*b*) 0.58
    (*c*) 0.62
    (*d*) 0.44.
11. Oscillating torque in ac machines causes
    (*a*) vibrations
    (*b*) noise
    (*c*) hunting
    (*d*) crawling.
12. Three phase a.c. machines are preferred over single phase machines because:
    (*a*) they are cheaper
    (*b*) single phase machines do not produce a pulsating magnetic field
    (*c*) single phase machines do not produce a rotating magnetic field
    (*d*) none of the above.
13. The active and reactive powers through a line section are 1 p.u. and 0.75 p.u. respectively. The voltage at the sending end of the section is 1 p.u. The resistance of the line section is 0.4 p.u. The total power loss, in p.u. is
    (*a*) 0.125
    (*b*) 0.375
    (*c*) 0.625
    (*d*) 0.5.
14. The active power loss in various sections of a distribution feeder are 1, 1.1, 1.2 and 1.3 p.u. The reactive power is 1.5 p.u. The sensitivity factor is approximately
    (*a*) 1
    (*b*) 2
    (*c*) 3
    (*d*) 4.

REFERENCES

1. ABDUL RAHMAN, T.K., and JASMON, G.B.: *A new voltage stability index and load-flow technique for power systems analysis'*, *Int. J.Power Energy Syst.*, 1997, **17**, (I), pp. 28–37
2. STERLING, M.J.H., CHEBBO, A.M., and IRVING, M.R.: *'Reactive power dispatch incorporating voltage stability'*, *IEE Proc. C*, 1992,**139**, (3), pp. 253–260
3. JASMON, G.B., and LEE, L.H.C.C.: *'Maximizing voltage stability in distribution networks via loss minimization'*, *J. Electr. Power Energy Syst.*, 1991, **13**, (3), pp. 148 – 152
4. SALLAM, A., ABOUL-ELA, M.E., and ELARABY, E.E.: *'Monitoring and control of voltage stability in power systems'*. Proceedings of international conference on *Energy management and power delivery* (EMPD95), Singapore, 21-23 Nov. 1995, **pp.** 342–247
5. PAL, K.: *'Voltage stability conditions considering load characteristics'*,*IEEE Trans. Power Syst.*, 1992, **7**, (I), pp. 243 – 249WEEDY, H., and COX, B.R.: 'Voltage stability of radial power links', *IEE Proc.*, 1968, **115**, (4), pp. 528–536
6. LACHS, R., and SUTANTO, D.: *'Different types of voltage instability'*, *IEEE Trans. Power Syst.*, 1994, **9**, (2), pp. 1126–1 134
7. SUTANTO, D., and LACHS, W.R.: *'Radial voltage stability'*, Proceedings of IEEE international conference on *Advances in power system control, operation and management*, Hong Kong, Nov. 1991, pp.6348
8. STERLING, M.J.H., CHEBBO, **A.M.**, and IRVING, M.R.: *'Voltage collapse proximity indicator: behaviour and implications'*, *IEE Proc. C*, 1992. **139.** (31. vv. 241-252, ~ **I_I.**
9. JASMON, G:B., and LEE, L.H.C.C.: *'New contingency ranking technique Incorporating a voltage stability criterion'*, *IEE Proc. C*, 1993,
10. KASHEM, M.A., and MOGAWEMI, M.: *'Maximizing radial voltage Stability and load balancing via loss minimization in distribution Networks'*. Proceedings of international conference on *Energy management and power delivery* (EMPD '98), Singapore, 3-5 March 1998, pp.91-96
11. BARAN, M.E., and WU, F.F.: *'Network reconfiguration in distribution Systems for loss reduction and load balancing'*, *IEEE Trans. Power Deliv.*, 1989, **4**, (2), pp. 1401–1407
12. SARFI, R.J., and SALAMA, M.M.A.: *'Distribution system reconfiguration for loss reduction: an algorithm based on network partitioning theory'*, *IEEE Trans. Power Syst.*, 1996, **11**, (I), pp. 504–510
13. BARAN, M.E., and W, F.F.: *'Optimal capacitor placement on Radial distribution systems'*, *IEEE Trans. Power Deliv.*, 1989, **4**, (I), pp. 725-734 **140**, (2), pp. 87–90
14. IEE ***Proc.-Gener. Trunsm.*** *Distrih., Vol. 147,* ***No.*** *3, May 2000* **175**
15. IEE2000 paper on, *Network reconfiguration for enhancement of voltage stability in distribution network*, by M.A.KESHAM, V.GANAPATHY, G.B.JASMON.
16. Proceeding on IEEE TENCON02, *Reconfiguration and capacitor placement for loss reduction of distribution system*, by ZENG RONG, PAN XIYUYAN, XENG XINFU.
17. IEEE Transaction on power systems .Vol.12, No.2,May 1997, *Distribution Feeder reconfiguration for operation cost reduction.*

Chapter 7 FAULTS IN DISTRIBUTION FEEDER

7.1 INTRODUCTION

Rapid growth of electric power systems, during the past fifty years, has resulted in a large increase in the number of lines in operation and their total length. These lines experience faults due to various reasons. When a fault occurs on a distribution system, it is essential for system operators to isolate the faulted area and to restore as much load as possible in the out-of-service area through network reconfiguration without violating operation constraints.

The restoration of lines can be expedited if the location of a fault is either known or can be estimated with reasonable accuracy. Several methods of locating line faults have been developed to achieve this objective. The primitive method of fault location consisted of visual inspection. Other previously proposed approaches, for locating transmission line faults, consist of using voltages and currents measured at one terminal of the line or both terminals of the line. Methods have also been proposed in the past for estimating fault locations on radial transmission lines.

7.2 COMMON FAULTS IN FEEDER

Electric power distribution feeders are susceptible to faults caused by a variety of situations such as adverse weather conditions, animal contacts, equipment failure, accidents, etc. Distribution circuit faults result in a number of problems related to the reliability of service and customer power quality. In the past, the permanent interruption of customer service resulting from a blown fuse or are closer lockout was the only factor which was used to determine service reliability. However with the introduction of sensitive customer loads, such as adjustable-speed motor drives, which are sensitive to the voltage sags caused by faults, even temporary self-clearing faults have become an item of concern. To maintain a high-quality of electric service, it is essential for utilities to determine the locations and causes of faults and take effective measures to minimize the occurrence of future interruptions in service.

Power companies normally perform fault location and diagnosis based on the customers' outage reports. Upon receiving the trouble calls from the customers, the dispatchers look at the feeder configuration map and the protection design manual to determine the outage area. Then a crew has to be sent to patrol the outage area. The location and diagnosis of the fault in this manner can be a rigorous and time-consuming task. Temporary faults are especially difficult to identify since these do not often result in a blown fuse.

7.3 FAULT LOCATION

Previously proposed approaches for locating faults on Distribution lines can be divided into two categories.

- The method in the first category uses the high frequency components
- The methods in the second category use the fundamental frequency voltages and currents at the terminals of a line and parameters of the line and loads

Performance of the technique in situations where cables are used could also be an issue. Another technique has been proposed for estimating shunt faults on radial distribution lines based on the fundamental frequency components of voltages and currents measured at the line terminal. This technique does not consider 'laterals' in the distribution lines.

7.3.1 Overview of the Technique

The proposed technique is described by considering a fault on a distribution line whose single line diagram is shown in Fig. 7.1. The selected system consists of an equivalent source G and the line between nodes M and N. Loads are tapped at different nodes and conductors are of different types between any two nodes. The line includes laterals.

7.3.2 Data Acquisition

When a fault is detected, the fundamental frequency components of the pre-fault voltage and current phasors at node M are saved. These components experienced during the fault are estimated and the fault type is determined from the phasors observed after a preset time has elapsed after the inception of the fault. These actions are taken in real-time by the fault locating hardware.

7.3.3 Estimation of Fault Location

The pre-fault and fault data, along with line and load parameters obtained from a database, are used in an off-line mode to estimate the location of the fault. The computation consists of the following major steps:

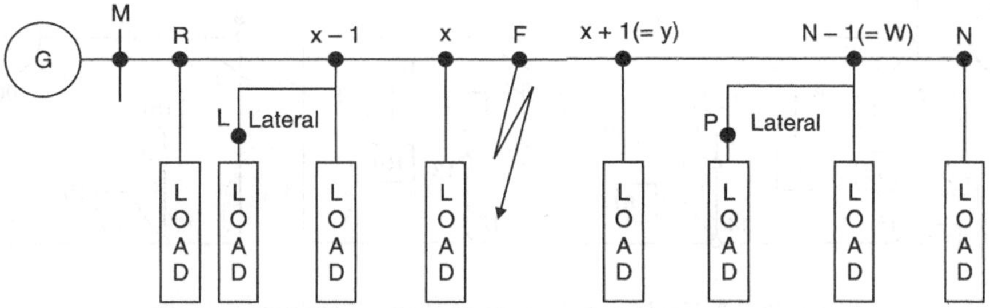

FIGURE 7.1* Single line diagram of a distribution line experiencing a fault at *F

1. **Estimating the faulted section.** The sequence voltages and currents at node M before and during the fault, are calculated from the estimated phases. A preliminary estimate of the location of the fault is made, let it be between nodes x and $x+1\,(=y)$. Line parameters, the type of fault and the phases of sequence voltages and currents are used to obtain this estimate. There could be multiple apparent locations depending on the 'laterals' on the line. Steps 2 to 5 are used for each apparent location.
2. **Modified radial system.** All laterals between node M and the apparent location of the fault are ignored and the loads of the laterals are lumped at the node from which the lateral is tapped. For example, load at tap L is lumped with the load at node $x-1$, for the fault path shown.
3. **Modelling of loads.** All loads up to node x are considered independently and the loads beyond the fault, node F, are assumed to be consolidated and are considered to be at the remote end, node N. Non-linearity of loads is modelled considering their dependency on voltage (8,9). The constants of load models, describing the voltage admittance relationship, are computed from the pre fault load voltages and currents.
4. **Estimating the sequence voltages and currents at the fault location and at the remote end.** Using the voltage dependent load model, determined sequence voltages and currents at node x during the fault are computed taking into account the load current at each node. The sequence voltages at the remote end are then calculated as a function of the distance of the fault from node x. The sequence voltages and currents at the fault node, F, are also obtained as functions of distance of the fault from node **x** and admittances of the consolidated load at the remote end.
5. **Estimating the distance of fault from node x.** Appropriate sequence voltage-current relationships at the fault node are used to iteratively estimate the distance of the fault from node x. The first estimate of the distance is obtained using the pre-fault sequence admittance of the consolidated load at node N. Subsequently, this admittance is updated using the new values of the sequence voltages at node N and voltage dependent load model determined. The procedure is repeated until a converged solution is obtained. Two additional estimates of the distance are obtained by considering that the fault is either located between the nodes x- 1 and x **01** between the nodes **x+l** and x + 2; if there is a reasonable chance of the fault being in those sections. Most plausible solution is selected and the location of the fault from the measurement node M is estimated.

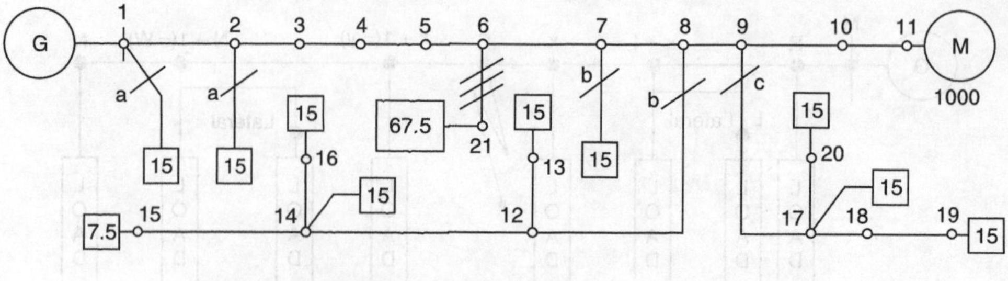

FIGURE 7.2 **Single line diagram of Power distribution circuit**

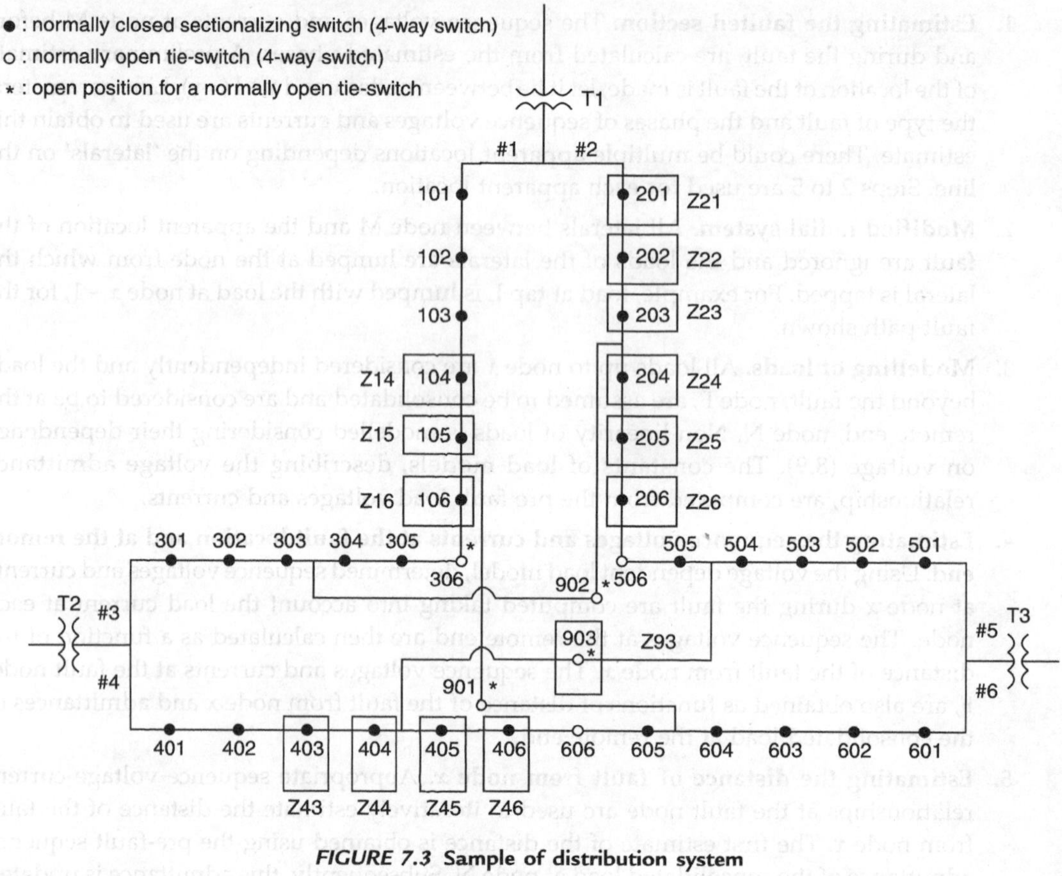

FIGURE 7.3 **Sample of distribution system**

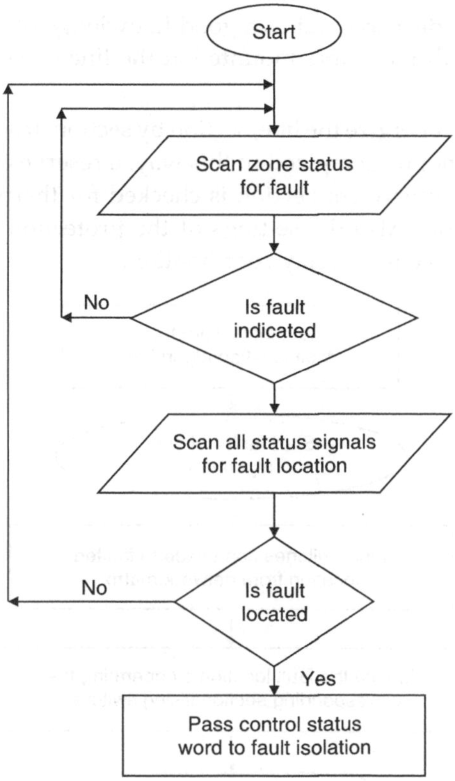

FIGURE 7.4 **Fault Location Algorithm**

7.4 FAULT ISOLATION

Once the fault location has been analyzed by the sub module, the automatic function for fault isolation and supply restoration is executed. The control of the network switches is done in such a way, that the line is processed zone by zone in the smallest possible remote controlled sections. After every reclosing of a line switch, the status of the feeding circuit breaker is checked, and if there has been tripping, the line section concerned is regarded as a faulty one.

The operation of the model is optimized so that the number of outages experienced by the customers connected to the faulty line is minimized. For this, there are two alternative approaches, depending on whether the faulty line section will be verified by trial switching or not. In the trial switching mode, the model first tries the line sections with the highest fault probability. When the faulty line section is encountered, it is isolated, and the remaining sections are energized. The trial switching mode actually performs a delayed auto reclosing. This mode is recommended in the cases, where there is some degree of uncertainty in the fault location, or where it is desired that the traditional delayed reclosing by protection relays be replaced by a zone-by-zone re closing.

In the second mode, the aim is to avoid the trial switching. Here the line sections are energized starting from the smallest fault probability and when only one section is left, the procedure is stopped. This mode is usually used only if a full reclosing procedure is first performed by the relays. To be effective, this mode also requires a good knowledge of fault location. In practice this means, that there must be fault indicators mounted at the line crossing points where the remote controlled switches are.

The model first tries to energize the line, section by section, from the initial feeding direction. For those sections which can not be energized in this way, a reserve connection is used. Before the coupling, the capacity of the reserve connection is checked for thermal load carrying ability and for the maximum voltage drop. Also the settings of the protection relays of the feeding circuit breaker are checked in order to ensure relay coordination.

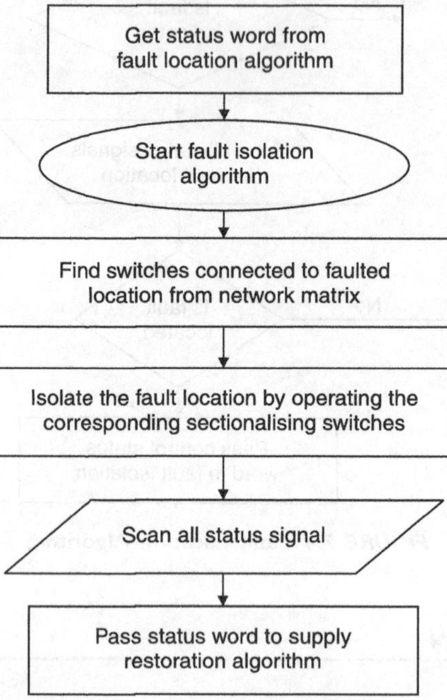

FIGURE 7.5 **Fault Isolation Algorithm**

7.5 RESTORATION

When a fault takes place in a distribution system, it is essential for system dispatchers to locate the fault, isolate the faulted zone, and restore the electricity service to the out-of-service area, which is outside the faulted zone. In recent years, a lot of works have been reported on fault location identification and service restoration. It is concerned with the problem of service restoration based on the assumption that the fault location has been identified and the faulted zone has been isolated.

7.5.1 Definition

Restoration is energizing the de-energized load by transferring them to the adjacent feeders by using tie switches. By restoration services interruption duration can be reduced significantly. Service restoration after a fault is known as emergency service restoration.

There are three methods of restoration,

1. Group restoration
2. Zone restoration
3. Load transfer.

The restoration algorithm is as follows

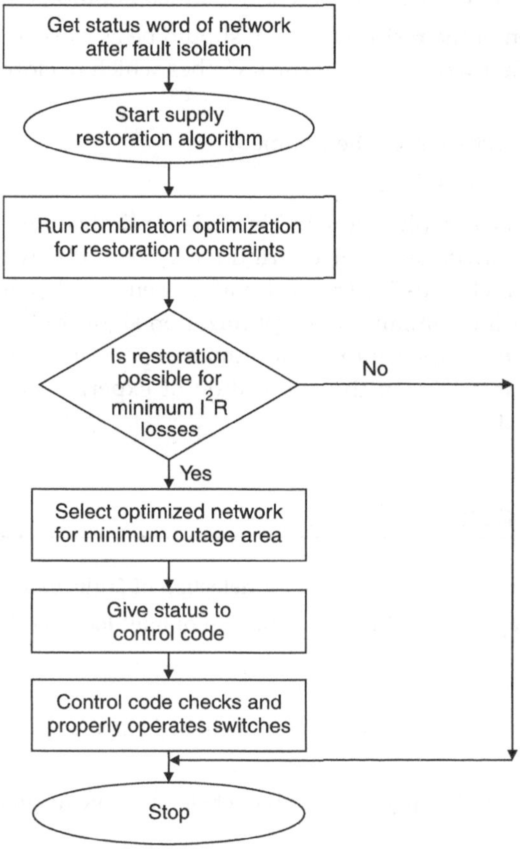

FIGURE 7.6 **Supply restoration algorithm**

The purpose is to determine a restoration plan for a fault on distribution feeders. Since the restoration plan reached in this work will be provided to system operators to help them make the final decision on how to restore the out-of-service area, it must meet the practical needs of system dispatchers as summarized below.

1. The restoration plan must be reached in a very short time. This is based on the practical consideration that the electricity service at the out-of-service area must be restored as soon as possible in order to avoid customers' inconvenience and enhance service reliability.

2. Restore as much load as possible within the out-of-service area.
3. The required number of switching operations in the restoration plan should be minimal. The main reason for allowing only minimal switching operations is that frequent operations may reduce the life expectancy of switches. In addition, it takes time for trouble men to operate a switch. This is especially a crucial factor in a crowded city because it may take the trouble men more than one hour to get to a place only one mile away. Therefore, the number of switches to be operated should be limited.
4. The configuration of the restored system should be as close to the original configuration as possible. In other words, only those switches which re close to the out-of-service area may be operated.
5. Radial system structure must be retained.
6. No components are overloaded.

Service restoration is a complicated combinatorial optimization problem because there are a great number of candidate switches in the distribution system. It may take a long time to reach a feasible restoration plan which satisfies the entire aforementioned practical requirements if one tries to solve the problem using combinatorial optimization algorithms. Therefore, the dispatchers at many utilities including the important role of human experts in reaching a suitable restoration plan motivated some recent works on the application of expert system approach and heuristic search approach to this field.

LONG TYPE QUESTIONS

1. Discuss the methods that were earlier used for detection of faults in a distribution system.
2. What are the major steps involved in the estimation of fault location? Illustrate your answer with a neat flow diagram.
3. What is fault isolation?
4. Enumerate the steps involved in fault isolation with the help of a flow diagram.
5. What is restoration?
6. Discuss the supply restoration algorithm with the help of a flow diagram.

MULTIPLE CHOICE QUESTIONS

1. In electric power distribution feeders, faults can be caused by
   (*a*) bad weather conditions (*b*) animal contacts
   (*c*) failure of equipment (*d*) all of the above.
2. Power companies perform fault location and diagnosis based on
   (*a*) Data acquired (*b*) Voltage profile
   (*c*) Customer outage reports (*d*) Past performance.

3. During fault location, if a fault is indicated
   (*a*) the zone status should be scanned for faults
   (*b*) all status signals must be scanned for fault location
   (*c*) check if fault is located
   (*d*) control status word for fault isolation must be passed.
4. For the successful of the fault isolation algorithm
   (*a*) we must have the status word from the fault location algorithm
   (*b*) we must find the switches connected to the faulted location
   (*c*) the fault location must be isolated
   (*d*) all status signals must have been scanned.
5. In the restoration algorithm, if it is not possible to restore the supply for minimum I^2R losses
   (*a*) the optimized network for minimum outage area must be selected
   (*b*) Status to control code must be given
   (*c*) Combinatorial optimization technique must be employed
   (*d*) We must stop.
6. Service restoration is a complicated combinatorial optimization problem because
   (*a*) there are many switches in the system (*b*) there are very few switches in the system
   (*c*) the systems are generally radial in nature (*d*) none of the above.

REFERENCES

1. Lemuel L. Means the Boeing Company Huntsville "*Development Of A Fault Isolation Procedure*" 1996 proceedings annual reliability and maintainability symposium.
2. J.T. Peng, C.F. Chien and T.L.B. Tseng "*Rough set theory for data mining for fault diagnosis on distribution feeder*" IEE Proc.-Gener. Transm. Distrib., Vol. 151, No. 6, November 2004.
3. R. Apel, C. Jaborowicz, R. Kussel "*Fault Management in Electrical Distribution Networks*" CIRED2001, 18–21 June 2001, Conference Publication No. 482 0 IEE 2001
4. Wen-Hui Chen, *Member, IEEE,* Men-Shen Tsai, *Member, IEEE,* and Hsi-Lin Kuo "*Distribution System Restoration Using the Hybrid Fuzzy-Grey Method*" IEEE TRANSACTIONS ON POWER SYSTEMS, VOL. 20, NO. 1, FEBRUARY 2005.
5. Final report of the CIRED Working Group WG03 Fault Management "*Fault Management in Electrical Distribution Systems*" This text comprises the final report of CIRED Working Group 03 "Fault Management". The working group was established in November 1995. The initiative came from Mr. Yves Harmand of Electricity de France, who was the Convener of the predecessor working group "*Distribution Automation*".
6. M S Sachdev R Das T S Sidhu University of Saskatchewan, CANADA

   "*Determining Locations of Faults in Distribution Systems*" Developments in Power System Protection, 25-27th March 1997, Conference Publication No. 434, *0* IEE, 1997

7. Jun Zhu, Member, IEEE, David L. Lubkeman, Senior Member, IEEE, Adly A. Girgis, Jun Zhu, Member, IEEE Fellow, IEEE *"Automated Fault Location and Diagnosis on Electric Power Distribution Feeders"* IEEE Transactions on Power Delivery, Vol. 12, No. 2, April 1997
8. J. A. Momoh, L. G. Dias and T. Thor,D. N. Laird *"Fault Simulation Studies of Single-Line Faults in a Delta-Delta Connected Distribution System"* CH 3381-1/93/$01.00 91993 IEEE
9. Chia-Hung Lin, *"Fault Restoration of Distribution Systems by Object-Oriented Programming"* Department of Electrical Engineering Kao Yuan Institute of Technology Kaohsiung, Taiwan
10. Pekka Verho, An Nikander, Pertti Jarventausta *"Aspects on Development of Distribution Network Fault Location and Management"* 2004 The Institution of Electrical Engineers. Printed and published by the IEE, Michael Faraday House, Six Hills Way, Stevenage, SGI 2AY

Chapter 8 REMOTE METERING

8.1 BACKGROUND FOR AUTOMATIC METER READING (AMR) FOR UTILITY

The traditional utility meter displays energy usage as an accumulation of counts presented to a display, which is used to calculate the monthly energy bill. It has applications within the electricity, gas and water utility industries for domestic, commercial and industrial applications. Sub metering is often used within a building, retail or industrial facility where it is desirable to measure energy usage for specific equipment, locations or sub-level accounts.

With rising energy costs and environmental pressures through the Climate Change Levy, there is a growing need to reduce energy consumption. Reducing losses and waste, and adopting energy efficient products and technology can achieve this.

Predicting energy usage remains a key issue in an industry where downtime is unacceptable. However, since energy losses are not easy to identify, it is difficult to determine the key culprits whether they are human users or power-hungry equipment. In many cases it is difficult to store energy *e.g.*, electricity, which must be generated to fulfil immediate demand. With rising energy costs there is an increased need to identify utility usage and waste and to take steps in reducing consumption through a remote Automatic Meter Reading (AMR) System.

8.1.1 What is AMR?

Automatic meter reading is the collection at a remote central location of data from meters and other devices at customers' premises via telecommunications. It involves the following components:

- Meter
- Meter register or index capable of generating pulses corresponding to the consumption through the meter, or creating an electronic data stream containing its current reading as well as additional information (cumulative consumption, peak demand, alarm flags, etc.)
- Telemetry interface unit (TIU) connected to the meter that transmits the information.

- Communication network or system to transfer the data from the TIU to the utility's offices. For radio systems, there may also be a local data collection unit (DCU) that gathers data from many nearby TIUs and transmits it over the communications network to the utility's offices.
- An AMR control computer or utility terminal unit (UTU) to receive, collect and manage this data.
- Software to run the system and present the data to the utility's billing and other information systems.

8.1.2 Benefits Influencing the Business Case for Automatic Meter Reading (AMR) Adoption

AMR remotely reads customer meters and then transfers the data into the billing system. AMR will reduce the need for meter readers to manually gather utility meter readings each month. Many utilities are using AMR as a way to improve customer service and control their meter reading costs, especially in areas with fenced yards, dogs, and landscaping and other issues that make accessing meters difficult or unsafe. Benefits include:

- Improved customer service, which includes:
- Minimizing the need to access customer property to read meters
- Reducing customer complaints and damage claims resulting from monthly visits to customer site
- Call resolution improvement - billing complaint calls will be handled more quickly due to availability of more frequent meter readings
- No need for customer to read their own meters due to meter access issues
- Provide detailed consumption information to customers
- Save on regular "on-cycle" meter reading and obtain more easily (and in some cases, instantly) initial and final meter readings for opening and closing accounts
- Streamline high bill investigations and flag potential high consumption before customers get a high bill
- Help pinpoint system losses and help detect theft of service
- Enable more frequent readings with little incremental cost
- Support customized rates (*e.g.*, for large customers) and improve cash flow management
- Improve revenue forecasting, by tying detailed consumption information to production data and expected billings
- Enable more sophisticated rates; *i.e.*, that more closely track costs, or encourage conservation.
- Improve meter management: ensure meters are properly sized; determine optimal time in service or cumulative registration
- Automatic Meter Reading (AMR) enables utility energy usage costs to be apportioned to location, equipment or sub-level accounts
- Fewer employee injuries, especially in areas with fenced yards, dogs and landscaping
- Improved billing accuracy
- Identification of utility power hungry equipment and in-efficient usage practices through comparisons of peaks and trends across sites and equipment through AMR.
- AMR telemetry provides feedback on utility energy saving measures where equipment usage translates into lost and saved cash and facilitates continual improvements for seasonal variations

8.2 COMPONENTS OF AMR SYSTEMS

8.2.1 Meter Interface Module

With power supply, meter sensors, controlling electronics and a communication interface it allows data to be transmitted from this remote device to a central location. In many instances, this communication interface is bi-directional and allows central office signals to be received by the remote unit as well. Every electric meter must have such an interface unit to be remotely read. Some key components of the remote device may be shared by more than one meter without regard for the type of meter; *i.e.*, electric, gas or water.

8.2.2 Communication Systems

Used for the transmission, or telemetry, of data and control signals between the meter interface units and the central office. Typically, such communications take the form of telephone, power line carrier (PLC), radio frequency (RF), or cable television. The system components in the communications system depend on the communication media used.

8.2.3 Central Office Systems / Equipment

Includes modems, receivers, data concentrators, controllers, host upload links, and host computer. Many utilities have for some time been taking advantage of electronic meter reading systems using hand-held data terminals that communicate with a central controller via phone lines. There is great similarity between the host side electronic meter reading and automatic meter reading system function. There are three major building block functions that the meter interfaces and related electronics must perform. These are common to electric, gas and water implementations.

The first functional building block, an electromechanical or electro-optical interface must be incorporated into or attached to the meter. This converts information conveyed by the meter's mechanical register indexes, or dial readings, into electronic signals that may be processed, manipulated, stored and transmitted.

The second functional building block is a controller unit consisting of a low-voltage power supply, signal processing electronics, microcomputer, and random access memory and program memory used to store the real-time run or operating system program. The controller unit is used to process the signals originating from the meter's electromechanical or electro-optical interface device. In effect, the controller unit converts the meter's electromechanical interface device signals into computer type electronic digital representations of the meter's exact index or dial readings-much as a calculator converts keypad entries into numbers appearing on the display. The controller's RAM memory maintains an up-to-the-minute mirror image of the meter's dials and as the dials increment, so do the numerical representations stored in RAM.

The third functional building block is the communication scheme and its associated transmit/ receive electronics. Generally, meter-to-utility host communications use one or more transmission techniques: telephone, power line carrier, radio frequency through the airwaves, or television cable. There are many sub-categories of each of these communication forms having to do with data flow, modulation techniques, and distance from remote site to central station and data transmission rates.

The AMR system starts at the meter. Some means of translating readings from rotating meter dials, or cyclometer style meter dials, into digital form is necessary in order to send digital metering data from the customer site to a central point. In most cases, the meter that is used in an

AMR system is the same ordinary meter used for manual reading. The internal mechanism used for metering consumption is identical in both cases. The one difference is the addition of some device to generate pulses relating to the amount of consumption monitored, or generate an electronic, digital code that translates to the actual reading on the meter dials.

8.3 COMMUNICATION METHODS USED FOR METER READING

8.3.1 Telephone Lines

Telephone lines are desirable from an economic point of view since most electricity users in the country have telephone service. The telephone system provides an ideal communication infrastructure for AMR systems as shown in Fig. 8.1 due to simplicity of operation, quality of data, high noise immunity, reliability and low cost, both at the remote site and the central station.

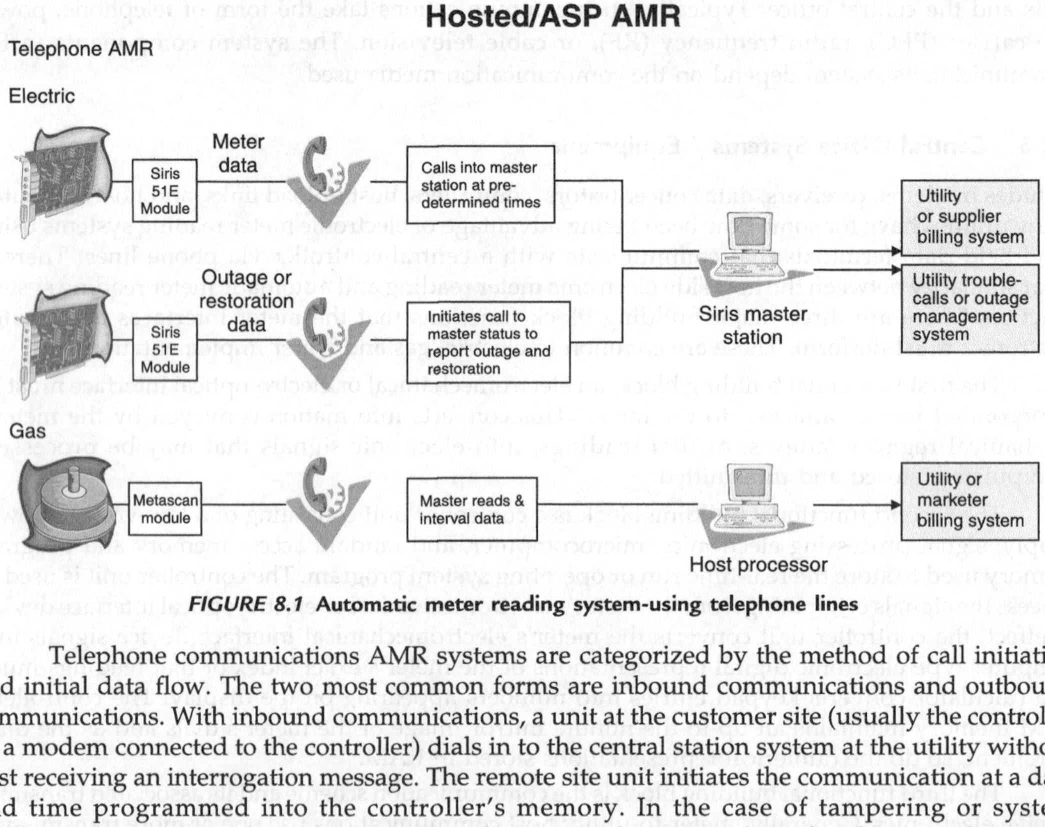

FIGURE 8.1 **Automatic meter reading system-using telephone lines**

Telephone communications AMR systems are categorized by the method of call initiation and initial data flow. The two most common forms are inbound communications and outbound communications. With inbound communications, a unit at the customer site (usually the controller or a modem connected to the controller) dials in to the central station system at the utility without first receiving an interrogation message. The remote site unit initiates the communication at a date and time programmed into the controller's memory. In the case of tampering or system malfunctions, a call can be initiated to the utility's central station, where the alarm condition will be received and processed. This approach takes advantage of the fixed monthly charge for local calls that the customer is already paying. No additional telephone access equipment is required.

The disadvantages of inbound communications are that the utility cannot obtain real-time data upon request, nor can the utility reprogram the controller unit or issue control commands as in the case of connect-disconnect or energy management, should these capabilities be incorporated into the system.

Outbound communications are those where data communications are initiated by a central unit located at the utility or at a local telephone company switching station. These systems respond to a query and require central telephone switching equipment and test trunk lines. Telephone Company involvement is required to enable the utility's central station computer to dial out to a customer's remote unit without ringing the customer's telephone. The advantage of this approach is that these systems function in real-time, as needed, which simplifies the implementation of demand load recording surveys, status monitoring, etc. The primary disadvantages to an outbound communications approach are the capital costs associated with the telephone company's involvement and the recurring tariffs that telephone companies charge. An additional complication arises in geographical areas served by one electric utility and two or more telephone companies.

A third approach is termed bidirectional communications. In this case, communications are initiated from the remote site or the utility's central station. The advantages of both inbound and outbound communications are incorporated in this system design. In the majority of cases, the inbound function is used, thereby reducing telephone charges. Also, due to the decreased density of outbound traffic, telephone company switchgear and test trunk lines are minimized.

8.3.2 Radio Frequency

Radio frequency, or RF, systems make use of small low power RF transmitters or transceivers located at the controller. These may take advantage of licensed or unlicensed portions of the RF spectrum and the effective radiated power of the transmitter and the distances capable of being traversed will vary as a function of the frequency and power of the remote transmitters and the receiving strategies employed. A variety of system configurations have been field tested thus far. The most successful employs a mobile unit operated in a van that sends a wakeup and transmits command to the remote meter units in its range. The remote meter units pick up the signal and respond by sending back requested data to the van's computer for later uploading and billing. This system is commercially available for use with gas meters. A variation of this approach employs remote meter units that regularly transmit every few seconds and a small portable receiver connected to a hand-held data terminal. Two of the more exotic approaches (in 1992) involve use of a cellular telephone network system and satellite communications. The mobile receiver approach suffers the significant disadvantage of being effectively mono-directional; thus, communication cannot be initiated from the utility's central office. Therefore, systems of this type have limited function and relatively low feature/function cost ratios and are not well suited for use by electric utilities.

8.3.3 Cable Television Communication

This communication approach uses existing cable television lines to transmit data. Some tests have shown that this may be a cumbersome and expensive approach but some municipal utilities that own cable systems are undertaking this type of communication. Additionally, many installed cable systems are not configured to pass signals from the subscriber's site to a central facility. It is expensive to upgrade these systems with wide-band bidirectional amplifiers and subscriber interactive taps. Cable television should not be discounted, however, as a viable communications medium. Several municipal electric utilities have purchased their local cable companies and upgraded systems consistent with the needs of AMR. If these utilities sell AMR services to local gas and water utilities, this approach can prove very viable. Future advances in cable will include bi-directional digital signal transmission and much wider band width ultimately using fibre optics at which point cable will be an ideal communications medium.

8.4 AMR SYSTEM

The remote Automatic Meter Reading (AMR) System is a host driven, multi-level network system consisting of a Host Central Station (HCS), Data Concentrator Units (DCU) and Meter Interfacing Units (MIU), with built-in flexibility and expandability. Each HCS, while working independently, can also be integrated with an existing corporate information management system through software interface. With additional hardware and software support, the HCS can function as a workstation in an existing Local Area Network (LAN) and becomes a member of the entire system, or several HCS can be connected together to form a network of their own.

The block diagram of an Automatic Meter Reading (AMR) system is shown in Fig. 8.2.

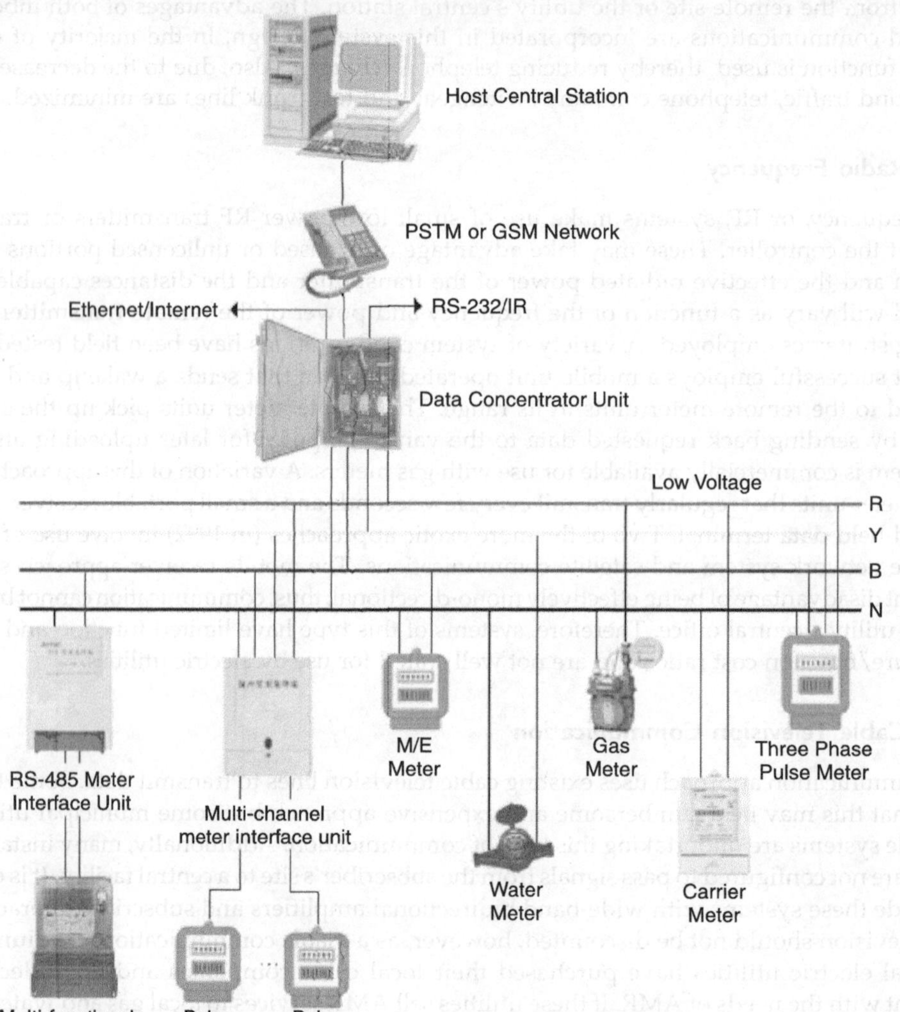

FIGURE 8.2 **Block diagram of an AMR system**

8.5 SERVICES AND FUNCTIONS

The full-scale implementation of AMR requires that a data communication network be established that effectively links every utility customer with the utility's central office. The actual amount of AMR-related data and its frequency of transmission is very low. These factors contribute to the difficulties encountered in the economic justification of AMR systems. There are, however, a myriad of services and functions that can be accomplished through this communication system, some of which significantly reduce a utility's operating costs and some of which can actually generate additional revenues. The incremental costs associated with incorporating these functions in the AMR system controllers are marginal. Payback can vary enormously. In theory, it is almost possible to finance a full-scale AMR system installation through the resulting costs savings and new revenue-producing services. Ten examples follow:

1. **Remote services connect and disconnect:** A recent survey in USA, indicated that the average annual turn-on and turn-off rate for electric utilities nationwide is 18 percent of total utility customers. Costs are incurred for sending a qualified service technician to a site to connect or disconnect a meter or power drop. With remote operation, these procedures are eliminated.
2. **Energy management:** Indirect load control and direct load control offer a financially attractive alternative to increased production and delivery capacity.
3. **Prepayment:** A number of utilities are investigating various prepayment systems. Such systems effectively eliminate collection problems and enhance cash flow.
4. **Combining utility services:** The incremental cost associated with manufacturing a controller capable of receiving input from three or more meters is marginal. By the same token, data can be retrieved at the host end and segregated and partitioned with relative ease and total security. By combining electric, gas and water AMR data on one system, the most expensive components—the communications infrastructure and the multitude of controllers-are shared by the utilities.
5. **Sub metering:** Multi-tenant properties can provide individual user data to the utility on a remote basis. The reduction of associated costs coupled with the magnitude of the transmitted information can, in many cases, allow the utility to offer its commercial customers new report-oriented services.
6. **Customer display:** Utilities can enhance customer relations by selling internal display units to customers that provide up-to-the-minute monitoring of power consumption, in rupees and paise. This can be used as an energy management tool and allows customers to verify and reconcile bills.
7. **Security system and services:** A utility so inclined could have a central office security system data communication infrastructure in place. The controller could serve as the communications gateway and the utility could operate a security business directly and use its monthly mailing of bills for marketing, or it could simply provide the data, on a contract basis, to an outside security company.
8. **Medical alert services:** As with security services, a utility could enter the rapidly expanding business of medical alert monitoring. The principles are the same as those for security services.

9. **Intelligent building applications:** An area of overlap between intelligent buildings and the AMR system is customer information coupled with the variable rate structures. The AMR system can act as a specialized interface between the building and the outside world. Information regarding operations, prevailing rates, customer usage, etc., would be instrumental to intelligent building owners and operators. This information would be made available to the building's computer and could be displayed on wall-mounted units as well. Such displays of rate information have been tested by utilities for years.
10. **Distribution automation and AM/FM:** Most utilities have begun computer-based mapping programs. By combining distribution automation and automated mapping systems and operating over the AMR network, any operational change occurring within the utility's equipment in the field can immediately be pinpointed on the computer based mapping system—a feature that greatly facilitates troubleshooting operations. When an outage occurs, data from the AMR and DA systems can be sent to the mapping system for immediate display as part of a geographical information map used for dispatching service personnel.

8.6 FINANCIAL ANALYSIS

Elements of a financial analysis of AMR include:

- Capital costs, including meters, TIUs, network, control computer hardware and software, installation, and project management
- Labour savings, including the reduction in staff hours for meter reading and meter reading-related customer service, and savings on vehicles and miscellaneous costs associated with meter reading
- Operation and maintenance of the system, including telecommunications costs, labour and vehicle costs, repair of system components, battery replacements
- Increased revenues from reducing system losses or replacing older inaccurate meters

The analysis should use discounted cash flow over a planning horizon, to account for life-cycle costs and residual value of the equipment.

Intangible benefits, even though they cannot be easily quantified, are important. They include improved customer service and confidence, and the value of consumption information to customers and the utility. In an era of privatization and increasing competition in the utility industries, these intangible benefits, and the customer loyalty they create, are increasingly important.

8.7 PLANNING FOR AMR IMPLEMENTATION

An AMR deployment is a high-profile project that requires careful planning. The implementation plan should include:

- Project control system
- Public and customer relations

- Project staffing
- Human resources plan for displaced employees
- Job descriptions for new responsibilities
- Customer service process redesign to take advantage of the new system.

8.8 SUMMARY

Automatic Meter Reading involves collection of data from meters and other devices at consumer's premises using telecommunications. It mainly consists of meter, an interface, the communication system and the software to run the system and present the information. The communication systems used include telephones, power line carrier, radio frequency (RF) and television cables, each of which have their advantages and disadvantages.

Implementation of AMR enables utilities to provide a host of other attractive services to the consumers. It leads to increased consumer satisfaction, accurate billing and transactions, investigation of system losses and thefts and provides feedback on energy management. Also, reduction in on-site meter reading costs and employee satisfaction and safety can be ensured.

However, all the financial aspects including capital costs, labour savings, O&M costs etc., should be suitably reconciled through Life Cycle Cost Analysis. Subsequently, proper planning should be undertaken before AMR is implemented as it is a high profile project, capital intensive and involves security and privacy issues.

LONG TYPE QUESTIONS

1. Explain the concept of automatic meter reading.
2. What are the benefits of automatic meter reading?
3. List the various components of AMR systems.
4. What are the various communication methods used for meter reading?
5. Explain the financial analysis of an AMR system.

MULTIPLE CHOICE QUESTIONS

1. In automatic meter reading, which of the following is connected to the meter that transmits the information?

   (*a*) Meter register (*b*) Telemetry interface unit
   (*c*) Utility terminal unit (*d*) Data collection unit.

2. The data collection unit is used to

   (*a*) Gather data from many nearby TIUs
   (*b*) Transmit the collected data to the offices of the utilities
   (*c*) Manage the collected data (*d*) Both (*a*) and (*b*)

3. An AMR control computer may also be called as

   (*a*) Utility terminal unit (*b*) Telemetry interface unit

   (*c*) Meter register (*d*) Data collection unit.

4. Telephone lines are a very popular communication media for meter reading because

   (*a*) They are simple (*b*) They have high noise immunity

   (*c*) They are economical (*d*) All of the above.

5. Consider the following statements

   (*i*) In inbound telephonic communication, the utility can obtain real time data upon request.

   (*ii*) In inbound telephonic communication, the utility can re-program the controller unit.

   (*iii*) Outbound telephonic systems cannot respond to a query.

   (*iv*) Outbound telephonic systems function in real time.

   Which of the above statements is true?

   (*a*) (*i*) only (*b*) (*ii*) and (*iv*)

   (*c*) (*i*) and (*iii*) (*d*) (*ii*) only.

REFERENCES

1. *Archnet Automatic Meter Reading (AMR) System Overview: - Archnet Electricity Meter Series*. Web site- http://www.archnetco.com
2. Cognyst Consulting LLC Automatic Meter Reading.
3. *Automatic Meter Reading* by Tom D. Tamarkin, published in Public Power magazine, Volume 50, Number 5, September-October 1992.
4. Denver Water- *Automatic Meter Reading*. Web site name - http://www.denverwater.org/custserve/residential/amr.html.
5. *Automatic Meter Reading* by Brian Lange & Morris Johnson, ESG (Energy systems group) - Rebuild Michigan Energy Services 2005 Technical Consultants Workshop Lansing, Michigan, January 13, 2005.
6. Colorado Springs Utilities, customer services- *Automatic Meter Reading*. Web site name - http://www.csu.org/customer/meter/page4795.html.
7. *All-Electronic Power and Energy Meters* By Paul Daigle, Product Manager, Analog Dialogue 33–2 (© 1999 Analog Devices)
8. US Meter Reading AMR Sites.
9. Two ways automated communication; web site- *www.twacs.com*, technical information.

Chapter 9 ENERGY AUDITING

9.1 INTRODUCTION

Electricity is the most flexible form of energy needed for all types of activity; hence due to this reason there is a rapid growth in the electric power demand from various categories of consumers. Electric power is considered to be the engine of the economic growth, especially in developing country like India with the progress of agriculture, industrial and commercial sectors largely depending on it. It is also a basic necessity for the domestic sector. Since much of the power generated is based on fossil fuels, it is a very precious commodity and should be used in an efficient manner. In the latest decades of twentieth century, India has been facing an acute crisis with the power generation not keeping up with the demand.

Energy audit is a tool to sense the opportunities to save energy. Energy audit plays a vital role in reducing the gap between the energy demand and supply. In a developing country like India, the energy audit with in utilities to enable to know the losses in various segments and devises packages to bring down the losses within the standard limits. So energy management and energy audit is very important for a developing country like India.

9.1.1 The Problems

The various problems associated with the increase in demand are

- Voltage Fluctuations
- Surges and Spikes
- Voltage Drops
- Unbalanced Voltage
- Poor Power Factor.

- Energy inefficient equipment.
- Inadequate motor loading.

9.1.2 The Symptoms

- Equipment and Switch Gear over Heating
- Frequent Shutdowns and Early Failure
- Reduction in Efficiency
- Power Loss
- Equipment Malfunction
- Power Factor Penalties
- Increased Conductor Losses
- Frequent use of Generators
- Production Loss
- Increased Maintenance Bill.

These factors may result in as much as 20% of energy losses and can even permanently damage your machines and equipment.

9.1.3 The Diagnosis – Energy Audit

The team of professional firm can conduct a detailed study of works. In this study, expert energy auditors will analyze the following:

- How much energy consumption of unit / factory
- Energy losses occurring in system
- Methods to curb the losses and save energy
- Techno-economic feasibility
- Improvement in safety and protection.

Thus, energy audit will help to make unit more energy-efficient, consequently saving energy and money. So, take the first step to improving company's bottom-line.

9.2 ENERGY AUDIT

Energy audit is the tool, to analyze and make step by step approach towards decision making for energy management. It helps to balance the total energy input with its use. Energy audit senses the opportunities to save energy.

An energy audit is a technical review of energy used by a building or process to determine where savings can be made. Just as a financial audit reviews the management and monitoring of money, an energy audit assesses the way energy is managed in an organization, and the procedures which control it.

An energy audit is not a substitute for a planned energy management programme which is integrated into an organization's operation. Energy auditing is only one element of an energy management information system. Without a supportive management and implementation structure, an energy audit is something of an orphan. You may have a good audit report but be unable to implement its findings effectively.

9.3 TYPES OF ENERGY AUDIT

The type of energy audit implementation depends upon:

- Type of industry and its function,
- Depth to which final audit is needed,
- The magnitude and potential of cost reduction desired.

Energy audit helps in sorting out the ways to reduce energy consumption per unit of product output or to reduce the cost of operation. The type of energy audit implementation should give energy conservation as a result.

The energy audit can be of two types:

1. Preliminary audit and 2. Detailed audit

9.3.1 Preliminary Audit

The preliminary audit alternatively called a simple audit, screening audit or walk-through audit, is the simplest and quickest type of audit. It involves minimal interviews with site operating personnel, a brief review of facility utility bills and other operating data, and a walk-through of the facility to become familiar with the building operation and identify glaring areas of energy waste or inefficiency.

Typically, only major problem areas will be uncovered during this type of audit. Corrective measures are briefly described, and quick estimates of implementation cost, potential operating cost savings, and simple payback periods are provided. This level of detail, while not sufficient for reaching a final decision on implementing proposed measures, is adequate to prioritize energy efficiency projects and determine the need for a more detailed audit.

Preliminary audit is worked out in limited time slot (5 to 12 days). In this, the study is carried out on energy supply and required demand. This study helps in giving status of progress of energy management if designed earlier.

9.3.2 Detailed Audit

The general audit alternatively called a detailed or mini-audit; site energy audit or complete site energy audit expands on the preliminary audit described above by collecting more detailed information about facility operation and performing a more detailed evaluation of energy conservation measures identified. Utility bills are collected for a 12 to 36 month period to allow the auditor to evaluate the facility's energy/demand rate structures, and energy usage profiles. Additional metering of specific energy-consuming systems is often performed to supplement utility data. In-depth interviews with facility operating personnel are conducted to provide a better understanding of major energy consuming systems as well as insight into variations in daily and annual energy consumption and demand.

The detailed audit gives correct analysis of conservation of energy and its cost. It formulates a detailed plan on the basis of quantitative and control evaluation. It accounts for 97% of energy utilized.

9.4 ENERGY AUDITING STEPS

Energy audit increases awareness of energy issues among people, which helps in increase in productivity and minimizes losses.

Energy auditing consists of the following steps:

1. *Planning and assigning responsibility:* In smaller operations, the business owner or facility manager may conduct the audit. For larger operations, an energy management committee may be necessary to perform the required work.
2. *Basic data collection:* Review of utility bills from the past year or from a typical twelve-month period. Also, look at data from accounting and production records to find out which areas of the facility or operations may be using the most energy. Record the current rate structure, demand and consumption figures, and any other fees for natural gas, electric, and other fuels.
3. *Data analysis*
4. *Formulation of low cost recommendation*
5. *Computation of payback period*
6. *Action plan is drawn up*
7. *Presentation of report*

Energy auditing program should cover points such as:

- Assigning the responsibilities of energy accounting, auditing by top management.
- Energy supply and production data.
- Priority-wise study of area requiring conservation.
- Implementation of action plan.

9.5 CASE STUDY

Savings Through Energy Efficient Lighting

Annual requirement

The number of 40 watt tube light average 10 hrs working is as under

Main office	–	114 Nos
Security block	–	10 Nos

Canteen	– 12 Nos
Old substation	– 25 Nos
Water plant	– 10 Nos
Generator shed	– 10 Nos
Boiler house	– 10 Nos
R & D Lab 1	– 90 Nos
R & D Lab 1	– 54 Nos
Q.A. Lab	– 36 Nos
Factory Street light	– 15 Nos
Colony	– 30 Nos
Total	– 418 Nos

This 40 watt tube light consumes additional 12 watt power for losses with total power consumption of 52 watt.

These can be replaced by energy efficient tube lights with only 28 watts power consumption.

(*a*) **Annual savings** $= \dfrac{((52 - 28) \times 418 \text{ nos} \times 10 \text{ hrs} \times 350 \text{ days} \times \text{Rs } 4)}{1000}$

$= \text{Rs. } 140448$

(*b*) **Investment.** Cost of energy efficient tube light = Rs. 800

Total cost of replacement = 418 × 800 = Rs 334400

(*c*) **Pay back period**

$$\text{Payback period} = \frac{\text{investment}}{\text{annual savings}}$$

$$= \frac{334400}{140448}$$

$$= 30 \text{ months.}$$

9.6 SCOPE OF ENERGY AUDIT

Energy Audit covers the following aspects in energy management

(*a*) It identifies energy loss that occurs in industry due to various reasons.

(*b*) It establishes quantum of energy loss.

(*c*) It suggests various methods of saving energy by improving network layout, installing energy saving devices, having energy efficient equipment and optimizing capacities of equipment.

(*d*) It determines the payback period by comparing money invested in energy saving equipment against money saved due to installation of energy saving equipment.

Improvement in safety and protection is also advised in the audit.

9.7 BENEFITS TO USERS

- Saving in Energy of 8 to 30%
- Saving of penalties like power factor penalty etc.
- Improvement in quality of products
- Increased Productivity
- Reduced cost of maintenance
- Fewer breakdowns
- Better safety and protection
- Identification of all points of energy loss
- Formulating strategies of energy and fuel saving
- Sharpening the competitive edge.

9.8 SUMMARY

For proper productivity, conservation and economical benefit, energy audit is compulsory. It is not a substitute for planned energy management programme, but only one element of energy management system. There are basically two types of energy audits: preliminary and detailed or general audit. The steps to be carried out for implementation of energy audit are covered and its beneficial role in a unit and its scope are detailed.

LONG TYPE QUESTIONS

1. Enlist the various problems associated with increase in demand and their common symptoms.
2. What is energy audit? Can it be a substitute for a planned energy management programme?
3. What are the types of energy audit?
4. What are the steps involved in an energy audit?
5. What are the benefits of energy audit?
6. The number of 40 W tubelights in various parts of an organization working 8 hours a day and 350 days in a year are:

   (*i*) Main office – 100
   (*ii*) Security block – 15
   (*iii*) Electronics lab – 90
   (*iv*) Machines lab – 50
   (*v*) Canteen – 12
   (*vi*) Corridors – 50
   (*vii*) Computer lab – 20

Assuming a 30% power loss in the ballast, calculate:

(*a*) The annual savings in the electricity bill.

(*b*) The total cost of replacement.

(*c*) The payback period.

Consider that the 40 W tubelights have to be replaced by 28 W energy efficient tubelights. The cost of each 40 W tubelight is Rs. 20 and each 28 W tubelight is Rs. 750. Ballast loss in 28 W tubelights can be neglected. The cost per unit of electricity is Rs. 7.5. [**Ans.** Rs. 263340, Rs. 252750, approx 1 yr]

MULTIPLE CHOICE QUESTIONS

1. In a company there are 80 fans, each consuming 100 W. If they are replaced by fans consuming 70 W each, the annual savings when they are operated for 10 hours a day for 350 days at Rs.4 per unit of electricity is

(*a*) Rs. 3600 (*b*) Rs. 45000

(*c*) Rs. 33600 (*d*) Rs. 20000.

2. In the above question, if each of the 70 W fans cost Rs. 1500, the total cost of replacement would be

(*a*) Rs. 150000 (*b*) Rs. 120000

(*c*) Rs. 300000 (*d*) Rs. 200000.

3. In the above question, the pay back period would be

(*a*) 43 months (*b*) 3 years

(*c*) 48 months (*d*) 1 year.

REFERENCES

1. T.W. Berrie "*Electricity Economy and Planning*", Peter Peregrinus ltd. on behalf of Institute of Engineers
2. A.S.Pabla "*Electric Power Distribution*" Tata McGraw hill companies. 4th edition, 2003.
3. Maimunah Ibrahim, M.Sc, MIEEE MOM Za" Jaafar, PhD Tenaga Nasional Berhad Malaysra "*DEMAND-SIDE MANAGEMENTMT* "IEEE *TENCON'93 / Beijng*, page no.572–576.
4. Roger Peters, Independent Expert, Canada. Assessment of Electricity Supply and Demand Management Options, *http://www.dams.org/*
5. Dr. D.N. Raut "*energy audit: a case study*", first national conference on Energy and fuel issues of future, November, 2004. Paper number 15.
6. Omar, A.; Mariun, N.; Mohd Amran Mohd Radzi "*Software Development for Energy auditing Practice*" Student Conference on Research and Development (SCOReD) Proceedings; 25–26 Aug. 2003 Page(s): 420 – 423
7. http://www.uet.edu.pk/ht3/energy%20auditing.html
8. www.mercindia.org.in/pdf/TPC_**Tariff_Booklet**.pdf
9. www.mercindia.org.in/pdf/BSES_**TARIFF_BOOKLET**.pdf

Chapter **10** *Energy Management*

10.1 INTRODUCTION

The power industry in India may seem to lack competition. This thought arises because each power company operates in a geographic region not served by other companies. Favourable electric rates are a compelling factor in the location of an industry, although this factor is much less important in times when costs are rising rapidly and rates charged for power are uncertain than in periods of stable economic conditions. Regulation of rates by State Electricity Boards, however, places constant pressure on companies to achieve maximum economy and earn a reasonable profit in the face of advancing costs of production.

"Economic Despatch" is the terminology referred to the process of apportioning the total load on a system between the various sources to achieve the greatest economy of operation. It is observed that all the plants in a system are controlled continuously as the load changes occur so that generation is allocated for most economical operation.

10.2 ENERGY MANAGEMENT

10.2.1 Types of Energy Management in Power Sector

Energy management is classified in power sector into three categories

1. Generation side management
2. Transmission side management
3. Demand side management

Generation side management deals with managing generating station for fuel requirement, maintenance scheduling etc., and thereby managing the generated power of a plant. *Transmission side management* deals with the maintenance of protective equipment, installation of new transmission line, power flow etc. *Demand side management* deals with the distribution of power, scheduling of load. In distribution system, we deal with the demand side management of a power system. Demand side management is also classified as:

(*a*) Need based energy management (*b*) Demand based energy management

Need based energy management forecasts the load according to the need of a consumer. Demand based energy management forecasts the load according to the demand for the load.

10.3 NEED BASED ENERGY MANAGEMENT (NBEM)

In power sector there is a distinct difference between demand and need. Consumers of electric power could be classified into five broad categories: the industrial users, agricultural sector, commercial organizations, domestic consumers and essential services. Industrial users could be further sub-divided as shift based industries and continuous process industries. Agricultural sector would include irrigation tube wells and rural industries. Out of these several groups and sub-groups, only three *viz* continuous process industries, domestic consumers and essential services need power round the clock, other may demand power for 24 hours of the day, but they don't need it.

Need Based Energy Management includes:

- Identifying the needs of various consumers.
- Forecasting the generation requirement based on the need.
- Plan power generation as per forecast.
- Lay down a suitable transmission and distribution network.
- Regulation distribution as per need.
- Monitor matching of need with supply.

With the advent of advanced Information Technology (IT) solutions, NBEM can refined and made to yield optimum results by drastically reducing transmission and distribution losses and enhancing consumer satisfaction. The IT tools designed to facilitate NBEM are SCADA Systems and Distribution Automation Systems.

10.3.1 The Distinct Advantages of NBEM

- It ensures high reliability of supply to consumers meeting the specific demand effectively for periods of actual requirements.
- The system losses can be substantially reduced since line and equipment does not get overloaded at any point of time.
- The voltage profile at all levels is improved thus safeguarding the customer's equipment from losing their efficiency and performance at low voltage.
- The scheme facilitates the adoption of energy conservation schemes and energy audit policy.

- Power cuts are reduced and quality of power improves leading to better industrial and agricultural health and productivity.
- NBEM can be successful only if there is political Will to go through the measures and administrative commitment to implement it faithfully. These are pre-requisites since vested interests are bound to be affected in any rational and scientific management system.

10.4 DEMAND SIDE MANAGEMENT (DSM)

Electricity end-use efficiency has the potential to play a key role in reducing the need for large dams. By managing electrical loads and maximizing the use of efficient electrical technologies, demand for electricity can be tempered and the need for large dams reduced. Many of these efficiency measures and technologies are cost effective at today's electricity prices and therefore provide real cost savings to energy users.

10.4.1 Definition

Demand-side management (DSM) programs consist of the planning, implementing, and monitoring activities of electric utilities that are designed to influence customer uses of electricity in ways that will produce desired changes in the utility's load shape.

Energy efficiency improvements provide multiple benefits:

1. ***Financial Benefits.*** Higher efficiency has direct and indirect financial benefits to Consumers and society by reducing the need for additional supply and/or distribution facilities, lowering customer energy costs, reducing equipment maintenance, creating opportunities for reducing equipment size, and mitigating risks from future price fluctuations.
2. ***Social Benefits.*** Efficiency investments can stimulate economic development and local job opportunities, improve energy security, and improve the quality of living and working conditions.
3. ***Environmental Benefits.*** Higher efficiency reduces both the direct emissions from On-site energy use, and the environmental impacts resulting from fossil fuel and electricity production and transmission.

While it is the public interest to use energy more efficiently, several market barriers exist which prevent energy users and utilities from making otherwise rational investments in efficient technologies and practices. For example, consumers often do not have access to new technologies, or cannot afford the extra cost involved or the financing that could pay for these costs out of the savings. Utilities often view efficiency in terms of loss of market share, and when trying to meet a growing demand tend to think first of new power supply options.

It has often been left to governments, therefore, to put policies and programs in place that encourages both consumers and utilities to undertake energy efficient investments and programs.

When an energy efficiency improvement program is carried out by a power utility or by a government or private sector organization as part of a strategic plan to manage electricity demand, it is called a Demand Side Management (DSM) program.

DSM is especially important to power planners in growing economies where installation of an inefficient infrastructure today means many years of wasted power.

10.4.2 Objective of DSM in Different Types of Economy

Following table shows a objective of DSM in different types of economy.

TABLE 10.1 Objective of DSM in different types of economy

Type of Economy	kWh per Capita	Examples	Objectives of DSM
Low Electricity Use	500	Parts of Africa and Asia	Manage peak demand and reduce losses
Rapidly Growing Economies	800 – 1400	China, India, and Latin America	Maximize efficiency and minimize peak demand in new equipment and facilities
Stable Industrialized Economies	7500	OECD	Meet new growth by increasing efficiency and load management in both existing and new stock
Economies with Uneven Growth		Parts of Asia and Africa	Limit peak demand in high growth sectors or regions
Economies with Older Inefficient Infrastructure		Eastern Europe	Upgrade current Infrastructure to high Efficiency

10.4.3 Unique Features of Demand Side Management

There are several unique features of power sector DSM which make both its evaluation and implementation very different from power supply options. These unique features present a challenge to power sector planners. There may be obvious country benefits from DSM (investment deferred, and environmental and social benefits etc.) but there may not be enough incentives for utilities to invest in DSM programs or consumers to invest in more efficient technology.

Features of Demand Side Management are as follows:

1. DSM involves actions and investments by energy consumers as well as Government and electrical utilities. DSM therefore has two main cost components—the cost of the new end-use technologies and practices themselves, plus the administrative and transaction costs of the program or policy to encourage (leverage) their use.
2. The benefits of DSM depend mostly on energy saved, and therefore they are strongly related to the costs and environmental and social impacts of the displaced energy supply source.
3. The size of the electricity and demand savings from a DSM program depends on the participation rate of energy consumers. This in turn depends on the regulatory or other methods used to overcome market barriers which might limit participation. The

sustainability of savings after a DSM program has been completed also depends on both the life of the end-use technology and the degree of market transformation achieved during the program.

10.4.4 Barriers in Demand Side Management

Compared with electricity supply options, DSM is disadvantaged by several market barriers——conditions which limit customer uptake of DSM measures or which reduce the incentive for electrical utilities to invest in DSM programs.

Barriers affecting customer uptake include lack of information and knowledge about energy efficiency, and financial considerations such as affordability, competing investment priorities, or access to financing.

Barriers preventing electrical utilities from undertaking DSM programs include lack of sufficient financial incentive because of deregulation and restructuring, hidden subsidies for other options, and lack of expertise and infrastructure to deliver DSM programs.

These barriers can be removed through appropriate government policy and regulation, and by careful design of DSM programs.

10.4.5 Solutions to Remove Barriers

(a) ***Lack of Information and Knowledge.*** For any product or service to be successful, consumers must be aware of its benefits, they must be able to evaluate it against alternatives, it must be widely available, and there must be sufficient local expertise (technical and managerial) to install and maintain the product.

Accelerated transformation of a market from conventional to efficient electricity using products and services therefore needs outside intervention in the initial stages.

Effective solutions include:

- Information and product promotion provided through channels well respected by industry, consumers, etc.
- Training programs for suppliers and users of energy efficient products and services in project audit, evaluation, design and implementation.
- Minimum Efficiency Standards covering the manufacture and sale of electricity using equipment.
- Labelling and endorsement of high efficiency electrical products and selective government procurement of these products.

(b) ***Financing/Affordability.*** There are several financial barriers which limit energy efficiency investments. Consumers with low credit ratings and low income communities can often not afford the first cost of energy-efficient equipment even when it pays for itself in the long run.

Financial institutions are often not familiar with the net financial benefits of energy efficiency investments.

Solutions that can remove some of these barriers include:

- *Innovative Financing Mechanisms:* Including product leasing in which lease payments are set equal to savings, performance contracting where energy service companies provide a "turn key" financing and implementation service, and revolving funds which can be used to finance for energy efficient projects.
- *Creation of a Level Playing Field:* For energy efficient and on-site renewable energy products by removing subsidies to conventional supply options, and moving to full environmental and social pricing of displaced energy supply sources through resource taxation or other methods.
- *Risk Sharing schemes:* Risk sharing through the use of Eco-Efficiency Funds secured by Government or industry associations.

10.4.6 Steps in DSM

DSM has the following steps

- Identify planning objectives
- Develop load curves and demand forecasts
- Identify supply and DSM options and collect data on each option
- Evaluate each option with respect to costs, benefits, size, life, and environmental, social, cultural, and economic development impacts
- Select most promising set of options
- Conduct scenario analysis
- Conduct risk analysis
- Conduct stakeholder view analysis with respect to externality values or weights
- Develop action plan

10.4.7 Technologies used in DSM

The energy conservation technologies are implemented to reduce total energy use. Specific technologies include energy-efficient lighting, appliances, and building equipment.

The technologies are categorized as follows:

10.4.7.1 Load Levelling

These technologies are used to smooth out the peaks and dips in energy demand - by reducing consumption at peak times ("peak clipping"), increasing it during off-peak times ("valley filling"), or shifting the load from peak to off-peak periods - to maximize use of efficient base load generation and reduce the need for spinning reserves.

Peak clipping, or the reduction of the system peak loads, embodies one of the classic forms of load management. Peak clipping is generally considered as the reduction of peak load by using direct load control. Direct load control is most commonly practiced by direct utility control of customer's appliances. While many utilities consider this as a means to reduce peaking capacity or capacity purchases and consider control only during the most probable days of system peak, direct load control can be used to reduce operating cost and dependence on critical fuels by economic dispatch. Another example of peak clipping is the use of interruptible/curtail able rates for industrial and commercial customers.

Valley filling is the second classic form of load management. Valley filling encompasses building off-peak loads. This may be particularly desirable for those times of the year where the long-run incremental cost is less than the average price of electricity. Adding properly priced off-peak load under those circumstances decreases the average cost to customers. Valley filling can be accomplished in several ways, one of the most popular of which is by adding new thermal energy storage (water heating and/or space heating) in place of loads served by fossil fuels (gas- or oil-fired).

Load shifting is the last classic form of load management. This involves shifting load from on-peak to off-peak periods. Popular applications include use of storage water heating, storage space heating, coolness storage, and customer load shifts. In this case, the load shift from storage devices involves displacing what would have been conventional appliances served by electricity (*e.g.*, installing thermal energy storage water heaters in place of regular electric water heaters).

Strategic conservation is the load-shape change that results from utility-stimulated programs directed at end-use consumption. The change reflects a modification of the load shape involving a reduction in sales often as well as a change in the pattern of use. In employing energy conservation, the utility planner must consider what conservation actions would occur naturally and then evaluate the cost effectiveness of possible intended utility programs to accelerate or stimulate those actions. Hence, the distinction between "naturally" occurring and deliberately induced changes in energy consumption and load shape is important. Examples include weatherization and appliance efficiency improvement.

Strategic load growth is the load-shape change that refers to a general increase in sales, stimulated by the utility, beyond the valley filling described previously. Load growth may involve increased market share of loads that are, or can be, served by competing fuels, as well as economic development in the service area. Examples include dual fuel heating, heat pumps, and promotional rates. In the future, load growth may include electrification. Electrification is the term currently being employed to describe the new emerging electric technologies surrounding electric vehicles, industrial process heating, and automation. These have a potential for increasing the electric energy intensity of the US industrial sector. This rise in intensity may be motivated by reduction in the use of fossil fuels and raw materials resulting in improved overall productivity.

Flexible load shape is a concept related to reliability, a planning constraint. Once the anticipated load shape, including demand-side activities, is forecast over the planning horizon, the power supply planner studies the final optimum supply-side options. Among the many criteria he uses is reliability. Load shape can be flexible if customers are presented with options as to the variations in quality of service that they are willing to allow in exchange for various incentives. The programs involved can be variations of interruptible or curtailable load; concepts of pooled, integrated energy management systems; or individual customer load control devices offering service constraints.

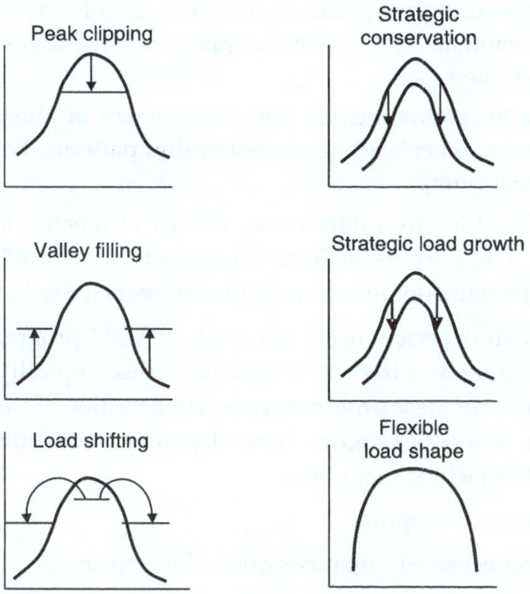

FIGURE 10.1 **DSM load shape curves (Demand in MW versus time)**

10.4.7.2 Load control

Energy management control systems (EMCSs) can be used to switch electrical equipment on or off for load leveling purposes. Some EMCSs enable direct off-site control (by the utility) of user equipment. Typically applied to heating, cooling, ventilation, and lighting loads, EMCSs can also be used to invoke on-site generators, thereby reducing peak demand for grid electricity. Energy storage devices located on the customer's side of the meter can be used to shift the timing of energy consumption.

10.4.8 Implementation of DSM

In recent research related to utility implementation of demand-side management programs, six different generic categories have been identified:

Customer education programs include utility programs designed to provide the customer with information regarding DSM options. Customer education on ways to conserve household energy consumption, or information on how a customer may qualify to participate in other utility DSM programs is example of such programs.

Direct customer contact programs usually involve some form of utility-customer interface (*e.g.*, energy audits, appliance service, etc.). Direct customer contact programs are particularly helpful in customizing and personalizing services to customers, and in responding to more technical, complex energy management concerns.

Trade ally cooperation programs involve working with organizations and groups that can influence the transactions between the utility and the customer. Examples include working with architects/engineers to promote energy-efficient building design, and cooperating with dealers to promote heat pumps or heat pump water heaters.

Advertising and promotion programs involve the use of various communication media, such as brochures, training seminars, etc., to provide appropriate messages to the customer regarding the technologies being promoted.

Alternative pricing programs include the development of alternative or innovative rate structures to influence the customer's energy consumption patterns. For example, off-peak rates may be used to promote heat pumps.

Direct incentives provided by utilities may consist of rebates, low-/no-interest loans, or other financial incentives. The purpose of financial incentives is to reduce barriers to customer acceptance by improving the return-on-investment to customers (or reducing the pay-back period).

Table 10.2 provides an overview of the six types of DSM programs. Electric utilities have used a number of different programs in these six generic types. Typically, successful utility efforts involve a combination of some of these program types. The selection of a particular implementation program, type of program, or mix of program types depends on the utility's overall objectives, as well as on a number of other factors, including:

- Type of technology or program
- Demographic and economic characteristics of the customer
- The utility's prior experience with similar programs and its overall philosophy
- Existing market penetration
- The receptivity of state regulatory authorities
- The estimated program benefits and costs to utility and customer
- Stage of customer acceptance
- Barriers to customer acceptance.

TABLE 10.2 Examples of implementation methods

Market implementation method	Illustrative objective	Examples
Customer education	• increase customer awareness of utility programs • increase perceived value of service	• bill inserts • brochures • information packets • displays • clearinghouses • direct mailings
Direct customer contact	• through face-to-face communication, encourage greater customer acceptance and response to utility programs.	• energy audits • direct installation • store fronts • workshops/energy clinics • exhibits/displays • inspection services

Trade ally cooperation (i.e.,architects, engineers, appliance dealers,heating /cooling contractors)	• increase utility capability in marketing and implementing programs. • obtain support and technical advice on customer adoption of demand-side technologies.	• cooperative advertising and marketing • training • certification • selected product sales service
Alternative pricing	• provide customers with pricing signals that reflect real economic costs and encourage the desired market response.	• demand rates • time-of-use rates • off-peak rates • seasonal rates • inverted rates • variable levels of service • promotional rates • conservation rates
Direct incentives	• reduce up-front purchase price and risk of low demand-side technologies to the customer. • increase short-term market penetration. • provide incentives to employees to promote demand-side programs.	• Low no-interest loans • cash grants • subsidized installation/modification • rebates • buy-back programs • rewards to employees for successful marketing of demand-side programs

DSM planning and implementation can be described in steps as shown in Fig. 10.2. The steps shown are only representative. The actual process is dynamic and would vary from utility to utility.

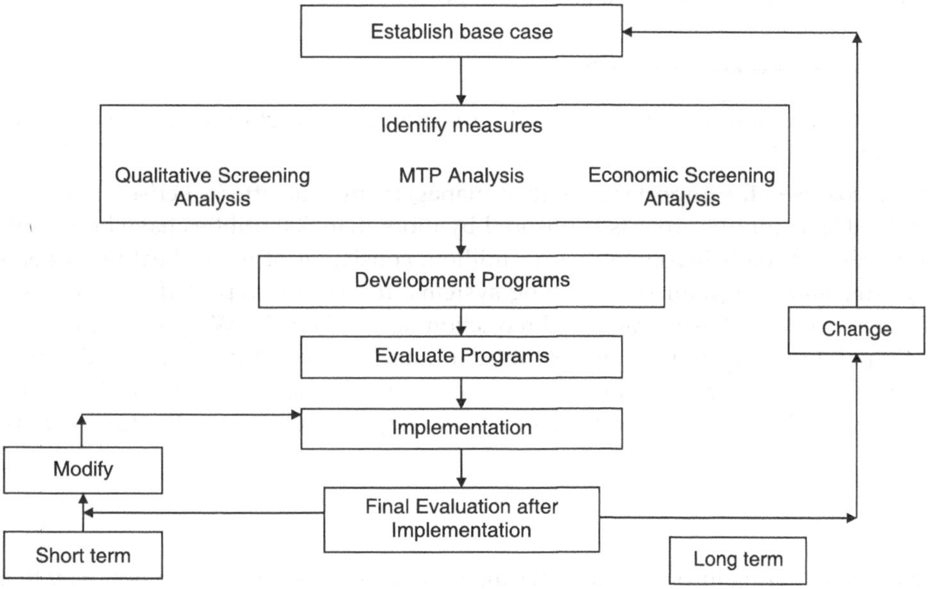

FIGURE 10.2 **DSM planning strategies**

10.4.9 DSM Economic Evaluation

DSM measures are expected to benefit both the customers and utility

A. Benefits to the utility

- Reduction in peak demand resulting in avoided capacity additions and costs
- Improved systems load factor thus reducing energy costs and efficient resource utilization
- Reduced risk of system collapse/load shedding and therefore more favorable system operation

B. Benefits to the customer

- Reduced power and energy costs offered as an incentive for co-operating with utility
- Reduced risk of unnecessary load shedding and system collapse

C. Benefit cost analysis

For the utility

$$\text{B/C Ratio} = \frac{\text{Avoided supply cost/kW}}{\text{DSM program costs/kW}}$$

Total DSM costs will have to be estimated on the basis of

Hardware costs+ Administration costs

For the customer

$$\text{B/C Ratio} = \frac{\text{Regular (power + energy) costs}}{\text{Reduced (power + energy) costs}}$$

Both costs are for the equal amount of power and energy.

10.4.10 International Experience with DSM

All the DSM measures are being planned and implemented, at least to some extent, in most of the countries of the world.

USA recognised the importance of load management soon after oil crises of 1973. Till 1990 more than 1750 load control projects sponsored by more than 500 utilities have been undertaken. The total number of loads involved was 13 million, consisting of water heating systems, space heating systems and central air conditioning systems. It has been reported that because of load control measures in USA, the average load reduction ranges from 1 KW/ point for winter heating to 28.7 KW/ point for irrigation pumps. Utilities in Canada, especially Quebec and Ontario, have also undertaken load management programs. Load management schemes have also been put into operation in Australia, New Zealand, Italy, France, England, Germany, Switzerland and many countries.

Tariffs

It is expected that offering incentives and pricing options to consumers is the best way to promote DSM. In USA, bill credits range from $1 to $5 per month for water heaters, $1.25 to $12 per month for air conditioners and \$15 to \$29 per HP per year for irrigation. In addition 7% to 50% rate reduction per kW demand, 8% to 20% reductions per kWh discounts are offered by many utilities. In UK, a tariff known as "Economy7" offers electricity during night time at less than half the day time rate. In Canada, DSM programs have resulted in savings of \$27 to \$48 per month. Similar incentives and multiple pricing options exist in a large number of countries in Europe and elsewhere.

10.5 SUMMARY

It is the electricity technology and the economics of the electricity industry which contribute to the volatility of price in a de regulated market. So, we should have sufficient supply to meet rising demand or reduce demand as per the available supply. This constitutes energy management, which is again of 2 types: need based and demand based. The former deals with connected loads of consumers which ultimately demands more and more generation for its growth and the latter deals with consumer's usage.

Demand Side Management (DSM) programs play a very important role in mitigating electrical system emergencies, avoiding blackouts and increasing system reliability, reducing dependency on expensive imports, avoiding high investments in generation and leading to environmental protection. DSM techniques are the cheapest, fastest and cleanest way to solve our electricity problems. These can be immediately implemented at $1/10^{th}$ the cost of building new power plants.

The basic premise of DSM is load levelling through peak clipping, valley filling, load shifting, flexible load shape etc. Incentives may be provided to the consumers for the same.

DSM economic evaluation is to be carried out through benefit-cost analysis or B/C Ratio.

LONG TYPE QUESTIONS

1. What is energy management? What are the various types of energy management?
2. How can the consumers of electrical power be classified?
3. What is need based energy management?
4. What are the advantages of need based energy management?
5. What is demand side management?
6. What are the advantages of demand side management?
7. State the features of demand side management.
8. What could be the possible barriers in adopting demand side management? Suggest a few solutions to remove these barriers.
9. Enlist the steps involved in DSM.
10. What are the technologies used in DSM?
11. What are the types of DSM implementation programmes?
12. On what factors does the selection of a DSM implementation programme depend?
13. Explain, using a flow diagram, the DSM planning strategies.

14. In a distribution system, for a generation of 500 kW, the cost of supply without applying DSM programme is Rs. 200000 and with applying a DSM programme is Rs. 100000. Calculate the B/C ratio for the utility.
15. For a load of 100 kW, the regular power costs Rs. 50000. On applying DSM, the cost is reduced by 75%. What is the B/C ratio?
16. The hardware cost of employing DSM is Rs. 65000, while the administration costs are Rs. 25000. If the B/C ratio is 3, what is the original cost of supply for a 500 kW generation?

MULTIPLE CHOICE QUESTIONS

1. Maintenance of protective equipment would fall under
   (*a*) Generation side management (*b*) Transmission side management
   (*c*) Demand side management (*d*) Need based energy management.
2. Which of the following falls under demand side management?
   (*a*) Generator maintenance (*b*) Installation of new transmission line
   (*c*) Power flow (*d*) Scheduling of load.
3. Which of the following statements is correct?
   (*a*) Demand side management is classified as generation side and transmission side management
   (*b*) Economic dispatch refers to the process of apportioning the total load of the system between different sources
   (*c*) System losses remain the same when NBEM is applied
   (*d*) None of the above.
4. In rapidly growing economies like India and China, the objectives of DSM would be to
   (*a*) Manage peak demand and reduce losses (*b*) Maximize efficiency and minimize peak demand
   (*c*) Limit peak demand (*d*) Upgrade current infrastructure.
5. In countries of Eastern Europe where infrastructure is inefficient, the major objectives of DSM would be
   (*a*) Manage peak demand and reduce losses (*b*) Maximize efficiency and minimize peak demand
   (*c*) Limit peak demand (*d*) Upgrade current infrastructure.
6. When an energy efficiency programme is carried out as a part of a strategy to manage electricity, it is called
   (*a*) Need based energy management (*b*) Demand side management
   (*c*) Environmental management (*d*) None of the above.
7. Which of the following statements is correct?
   (*a*) Demand side management has only one cost component
   (*b*) The benefits of DSM depend on the energy lost
   (*c*) DSM involves actions and investments by the utility alone
   (*d*) None of the above.

8. Load leveling technologies are used to
(*a*) Smooth out the peaks and dips in energy demand
(*b*) Reduce system peak loads
(*c*) Build off-peak loads
(*d*) Shift loads from on-peak to off-peak period.

9. Direct load control can be used for
(*a*) Reduction of system peak loads
(*b*) Reduction of capacity purchases
(*c*) Reduction of operating costs
(*d*) All of the above.

10. Valley filling can be accomplished by
(*a*) Direct load control
(*b*) By adding new thermal energy storage
(*c*) By increasing appliance efficiency
(*d*) By use of fossil fuels.

11. Strategic load growth can be defined as
(*a*) Change in load shape that refers to a general increase in sales
(*b*) Shifting load from on-peak to off-peak periods
(*c*) Change in the load shape resulting from utility simulated programmes
(*d*) None of the above.

12. Cash grants, no interest loans and rebates would be examples of
(*a*) Alternative pricing
(*b*) Direct incentives
(*c*) Direct customer contact
(*d*) Customer education.

13. The B/C ratio for the customer side is 3.5. If the regular power cost is Rs. 20000, the cost when DSM is employed is
(*a*) Rs. 4174
(*b*) Rs. 5417
(*c*) Rs. 5714
(*d*) Rs. 1754.

REFERENCES

1. T.W. Berrie "*Electricity Economy and Planning*", Peter Peregrinus ltd. on behalf of Institute of Engineers
2. A.S.Pabla "*Electric Power Distribution*" Tata McGraw hill companies.4th edition, 2003.
3. *Generation of Electrical Energy*, by B.R. Gupta, Eurasia Publishing House, Delhi, 4th Edition, 2003.
4. Maimunah Ibrahim, M.Sc, MIEEE MOM Za" Jaafar, PhD Tenaga Nasional Berhad Malaysra "*DEMAND-SIDE MANAGEMENTMT* "IEEE *TENCON'93 / Beijng*, page no.572-576.
5. Roger Peters, Independent Expert, Canada. *Assessment of Electricity Supply and Demand Management Options, http://www.dams.org/*
6. Dr. D.N. Raut "*energy audit: a case study, first national conference on Energy and fuel issues of future*, November, 2004. Paper number 15
7. Dr. M.K.Khedkar, "*Need based energy management- Adding a new dimension to EMS*" STTP on Demand Side Management at Walchand College, Sangli, July 2001.
8. http://www.cc.state.az.us/utility/electric/DSM-Exhibit1.pdf

9. Reynolds M. Delgado, *Demand-Side Management Alternatives*, Proceedings of the IEEE, VOL. 73, NO.10, October 1985, pp.1471–1488

10. Ashok M. Ghatpande, *Frequency Dependent Load Chipping as a Demand Side Management (DSM) Tool - Concept And Issues*, IEEE/IAS International Conference on Industrial Automation and Control, 5-7 January 1995, pp.383–388

11. C. Nunn, *Customer Communications For Demand-Side Management*, IEE Colloquium on DSM and Resource Planning in the UK and Europe(Digest No.1994/186),13th October 1994, pp. 3/1– 3/5

12. Ikhupuleng Dube, Musara Beta and Steven Dihwa, *Opportunities for Demand Side Management (DSM) in Integrated Energy Resource Planning and Development in the Power Sector: The Case of Zimbabwe*, Inaugural IEEE PES 2005 Conference and Exposition in Africa Durban, South Africa,11–15 July 2005, pp.234–240

13. Dr. G. Thomas Bellarmine, *Load Management Techniques*, Proceedings of the IEEE, SoutheastCon 2000, 7–9 April 2000, pp.139–145

14. Clark W.Gellings, *The Concept of Demand-Side Management for Electric Utilities,* Proceedings of the IEEE, VOL. 73, NO. 10, October 1985, pp.1468-1470

15. Dilip R. Limaye, *Implementation of Demand-Side Management Programs*, Proceedings of the IEEE, VOL. 73, NO. 10, October 1985 , pp.1503–1513.

16. Anibal T. De Almeida and Edward L. Vine, *Advanced Monitoring Technologies For The Evaluation Of Demand Side Management Programs*, IEEE Transactions on Power Systems, VOL. 9, NO. 3, August 1994, pp.1691–1697.

Chapter 11

Economics of Power Distribution

11.1 INTRODUCTION

For the proper management of any electricity supply undertaking, it is important that reliable methods of costing are used to run it on a sound financial basis. At a time when increasing attention is being paid to the conservation and efficient utilization of energy resources, any approach towards the economic optimization of the electricity supply system must surely accept that it is imperative to have a realistic costing basis. There are costs associated with owning, operating and maintaining the power distribution system. A suitable pricing and charging structure is required to achieve the economy in power distribution. The cost and performance of electricity utilities are influenced by many factors, such as geography and topography of the area, the number, nature and density of customers, the lengths and types of distribution circuits, weather conditions, asset replacement cycles, inherited network characteristics (pre-vesting investment) etc., which may be unique for each particular utility.

11.2 COST

'Cost' is the total sacrifice that must be expended or traded in order to gain some desired product or end result. It can include money, labour, material, resources, real estate, effort, lost opportunity, and any thing else that is given up to gain the desired end. Usually such a combination of many different resources and commodities is measured on a common basis 'money' by converting materials, equipment, land, labour, taxes and permits, maintenance, insurance, pollution abatement, and lost opportunity costs to the currency. In case where all the elements of cost can be put on to a common basis, the subsequent planning can be done in a single attribute manner, the goal being to minimize the monetary cost and the cost reduction basically trading the cost of one item against another to find the best overall mix. The following are different types of costs.

11.2.1 Capital Costs (CC)

Apart from the capital costs of owning the network assets, the O & M (operation & maintenance) costs associated with using the assets and the overhead costs of operating the business, there is also an ongoing need to invest capital into the existing asset base. The nature of electricity distribution and, more specifically the complex and high-tech plant and equipment required to provide an efficient and effective transportation network, results in the business being highly capital intensive. The age, condition, future requirements and increasing performance expectation means that, when compared to other commercial and industrial firms, distribution companies have relatively high capital investment requirements. The cost of substation, feeder, or other element of a power system can also be viewed as composed of fixed and variable costs. Fixed costs are those that do not vary as a function of any variable element of the plan or engineering.

For example, the annual cost of:

- taxes
- insurance
- inspection
- scheduled maintenance
- testing, etc. required to keep a 100MVA substation in service do not vary depending on its loading.

Variable costs include only those that change depending on its load or utilization: if more loads are transferred to it so that it serves a higher load, its losses go up; if low-side voltage is regulated more tightly, its LTC maintenance costs will probably increase. (Fig 11.1)

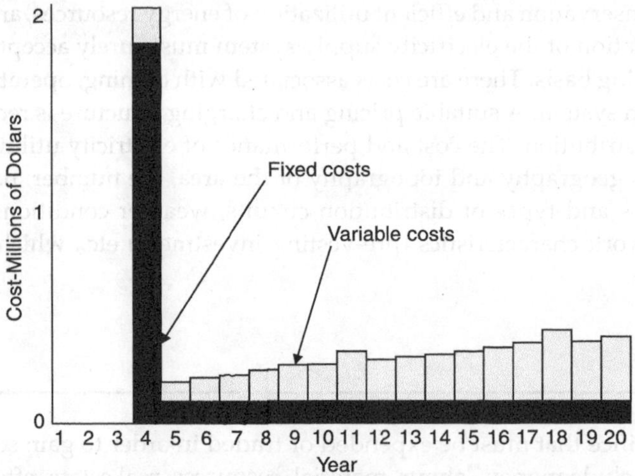

FIGURE 11.1 **Fixed and variable costs**

The new substation's costs can also be viewed as composed of both fixed costs—those that are constant regardless of loading and conditions of its use—and variable costs, which change depending on load, conditions, or other factors in its application as part of the power system. In this example, the variable costs increase nearly every year—reflecting higher losses costs as the load served gradually increases due to load growth.

11.2.2 Operational Costs (OC)

From the regulator's perspective, OC is the amount of money that is believed to be required to run an efficient distribution business. These are essentially expenses, or revenue, costs and are charged to the profit and loss account as operating costs.

They typically include

- Employee wages
- Office rent
- Premises energy bills
- Transport and fuel
- Billing etc.

Costs associated with the operation, repair and maintenance represent a significant element of OC and are required to keep the distribution system in proper working order throughout the year.

O & M costs provide for:

- Monitoring and controlling the system (*e.g.*, network switching operations)
- The implementation of safety rules and regulations
- The development and maintenance of an effective system repair and restoration service
- Asset maintenance.

Operational costs associated with supply availability are particularly significant in rural areas where customers are connected to parts of the distribution network, which are often both electrically weak and physically more vulnerable. Costs associated with repairing the network, especially those which are weather related, can often be mitigated by increased capital expenditure, *e.g.*, stronger wires and a greater penetration of automatic network switching.

11.2.3 Cost of Raising Capital

This represents the level of return required by the financial markets in order to attract sufficient investment into the business. It defines the interest, or additional funds generated as a result of investing money in the distribution system. This rate of return on investment is determined by the financial markets and represents the investor's expectations in terms of financial performance. Most distribution company capital is obtained from both bank loans (debt capital) and shareholder investment (equity capital). The cost of capital is usually calculated as a weighted average cost from each of these two sources. The objective is to develop a company financing structure which delivers the optimum weighted average cost of capital.

11.2.4 Embedded, Marginal, and Incremental Cost

Embedded cost is that portion of the cost that exists in the current planned system or configuration or level of use. Depending on the application, this can include all or portions of the initial fixed cost and all or parts of the variable costs. Marginal cost is the slope of the cost function at the current operating point, a point at which current embedded cost is defined. Incremental cost is the cost per unit of a specific jump or increment. Marginal cost and incremental costs both express the rate of change of cost with respect to the base variable, but they can differ substantially because of the discontinuities and non-linearities in the cost function.

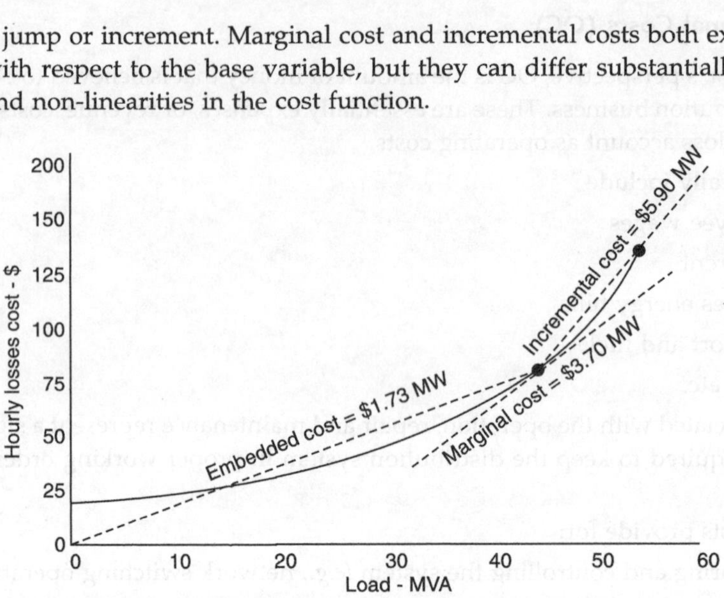

FIGURE 11.2 **Embedded, marginal and incremental costs**

Hourly cost of losses for power shipped through a substation transformer as a function of the load. Currently loaded to 45 MVA, it has an embedded losses cost of $1.73/MVA and a marginal cost of losses at that same point of $3.70/MVA. The incremental cost of losses for an increase in load to 55 MVA averages $5.90/MVA.

11.2.5 Return on Investment (ROI)

If a business is to secure access to funds on attractive terms then it must be capable of delivering a return on the invested capital. The level of return must be commensurate with the exposure to risk and, more importantly must reflect market expectation. Establishing the current value of the existing asset base is an important aspect in determining the levels of allowed revenue, for it is on this figure that a 'market rate of return' is allowed to be earned. This regulatory valuation of the capital worth of the distribution network is also used as the basis for calculating depreciation allowance which represents a significant proportion of the allowed distribution revenue.

11.2.6 Depreciation

The decline in the value of the network assets due to wear and tear, or technical obsolescence is an important business cost that must be allowed for, in order that it can be appropriately recovered. The purpose of depreciation in accounting terms is to gradually remove the cost of the asset from the accounting records by treating it in the profit and loss account as a cost of business operation. The most commonly used method is 'straight line' depreciation, which attributes an equal fraction of the original cost (minus any residual asset value) to each year of the asset's useful life.

11.3 REVENUE REQUIREMENTS

In order to operate as a business, an electric utility must take in sufficient revenue to cover the following important financial obligations:

- Continuing operating costs
- Pay for its equipment and system
- Cover its debt payments (loans and bonds) and
- Provide earnings for its owners (shareholders).

11.3.1 Breakeven Point

A planner's job is to balance factors mentioned above to determine how they contribute to decision of when to build an item or actually commit to an expense and to identify what alternative with regard to timing has the lowest overall cost (or, alternately, the highest benefit) to the company. The tariffs in public power utilities are generally fixed to breakeven costs. This means at breakeven point the fixed annual cost of capital equalizes the revenue earned over the financial period as shown in Fig. 11.3.

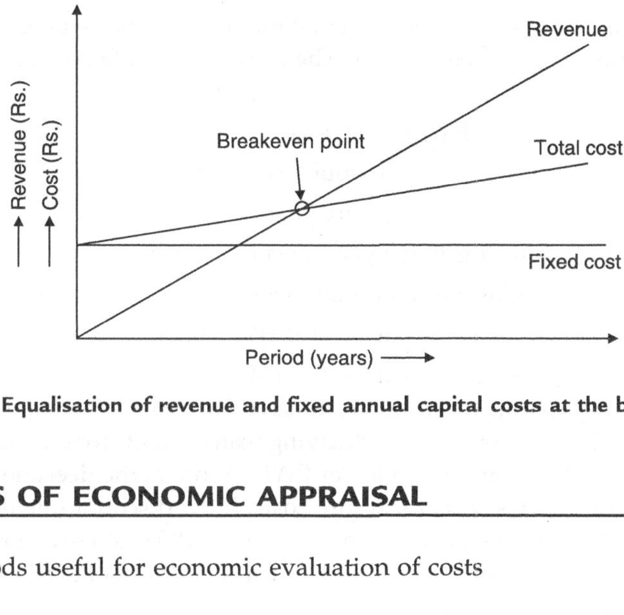

FIGURE 11.3 **Equalisation of revenue and fixed annual capital costs at the breakeven point**

11.4 METHODS OF ECONOMIC APPRAISAL

The following methods useful for economic evaluation of costs

11.4.1 Present Worth Analysis

Present worth analysis is used to evaluate and rank alternatives when there is a difference in the timing of expenses—the substation could be built this year or next year or when costs and savings that occur at different times must be balanced—more spent today will lower tomorrow's costs. It is an essential element of keeping cost as low as possible and it should be applied in all situations where alternatives differ with respect to when expenses will be incurred. The present worth factor is the tool which makes these comparisons valid. The factor represents the value of money a year from now in today's terms.

The value of money at any time in the future can be converted to its equivalent present worth (present value) as

Value today of X Rupees t years ahead = X * P^t

where P is the present worth factors

For example, suppose P=0.9, then Rs. 100 a year ahead of now is considered equal in value to today's money of,

$$Rs.100 * (0.9)^1 = Rs.\ 90$$

and five years from now is worth only Rs.100 * $(0.9)^5$= Rs.59.05

Alternately, the present worth factor can be used to determine how much future money equals any amount of current funds, for example, to equal Rs. 100 present worth, one year from now the utility will need

$$Rs.\ 100 / (0.9)^1 = Rs.111.11$$

and two years from now, one would need Rs.100 /$(0.9)^2$= Rs. 123.46 to equal the present worth of Rs. 100 of today.

Factors Affecting Present Worth Analysis

(a) *Discount rate:* The discount rate used in the analysis, d, is the perceived rate of reduction in value of money from year to year. The present worth factor is related to the discount rate as,

$$P(t) = 1/(1 + d)^t$$

where d = discount rate and

t = future years

e.g., If d is 11.11%, it means that a year ahead money is discounted 11.11% w.r.t. today's money, equivalent to present worth factor of,

$$P = 1/(1 + 0.111)^1 = 0.90$$

A decision making process based on the values of d = 11.11% and PWF = 0.90 would conclude that spending Rs. 90 today to save Rs. 100 a year from now is a breakeven proposition, while if the same Rs. 100 savings can be made for only Rs. 88, it has a positive value. A higher discount rate—a lower PWF—renders the decision making process less willing to trade today's costs for future costs. If the discount rate is doubled, to 22.22%, P drops to 0.82. Now, Rs. 88 spent to obtain a Rs. 100 cost reduction a year from now is no longer a good investment: that Rs. 88 must save at least Rs. 88 /1.222 = Rs. 108 a year ahead from now to break even.

(b) *Interest rate*: The fact that it takes more future money to equal the money today is often attributed to the interest rate : a person who has Rs. 100 today can invest it at the prevailing interest rate, i, so that after one year it will be $(1 + i)$ times as much. The rate depends on several factors including the growth indexes, economic policies, bank rates etc.

(c) *Earning targets:* The present worth factor for a Rupee under the utility's control should be better than what is obtained from the interest basis ($1/(1 + i)$), because the utility must be able to do better with its earnings than the prevailing interest rate.

For example, while the prevailing interest rate may be 5%, the utility's financial planners may desire an earning potential of 12% on all new expenditures. Rather than a PWF of 0.952 (5% interest) the utility would use a PWF of 0.892 (12% interest) to interject in to the decision making process its unwillingness to spend Rs. 100 today unless it returns at least Rs. 112 after one year from now.

(*d*) ***Financial risk and expected return:*** In general, interest rates reflect the level of risk that creditors see in a person, institution, or business model. A low element of risk means a relatively low interest rate, and vice versa. Ideally creditors adjust their interest rates to take into account the risk inherent in each loan or investment, with a goal of assuring that their expected return is positive.

(*e*) ***Inflation:*** Inflation is a factor that further discounts the value of money. Inflation is rarely used in planning the future expenses and benefits due to its uncertain nature governed by the balance between demand and supply of goods and services.

(*f*) ***Internal monetary risks:*** In practice, the present worth factor used by the utility, at least the implicit PWF that is really applied in its decisions, often includes a margin to account for internal risk such as unprecedented failures of equipments or process.

11.4.2 Life Time Cost Analysis

Levelised cost analysis is a particularly useful way of comparing plans when actual costs within each case vary greatly from year to year over a lengthy period of time. Often it is useful to compare projects or plans on the basis of their average annual cost, even if their actual cost changes greatly from year to the next. This involves finding a constant annual cost whose present worth equals the total present worth of the plan.

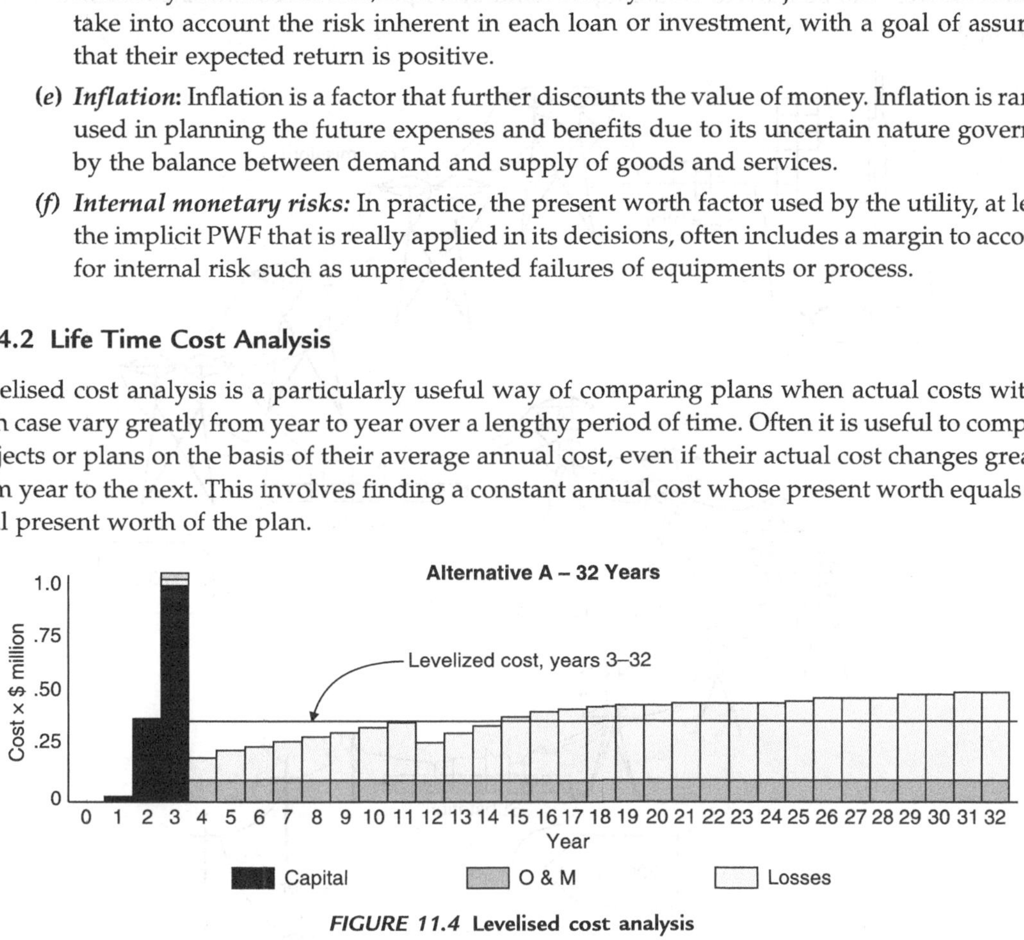

FIGURE 11.4 **Levelised cost analysis**

Present worth of all costs sums to slightly more than $3,000,000. Levelized over years three through 30 – those in which the substation is planned to be in service—this represents an annual cost of nearly $350,000 per year which can be viewed as the annual cost of providing the substation service.

For example, if a substation is to be constructed, it would be planned on the basis of providing a minimum of 30 years of service and it would be carried as a depreciating asset on the company's books. The figure above shows projected cost for Alternative A over 32 year period and it shows a levelized cost of Rs.1.742 Cr. per year. If an Alternative B exist that has a levelized cost of Rs. 1.711 Cr. per year, it will save Rs. 3.15 Lakhs per year in terms of 30 years budget perspective.

11.5 VARIATIONS IN COSTS

Ideally the cost of power distribution is fairly linear with respect to both the amount and the distance power is to be moved. Doubling the amount of power moved and the cost will double, moving it

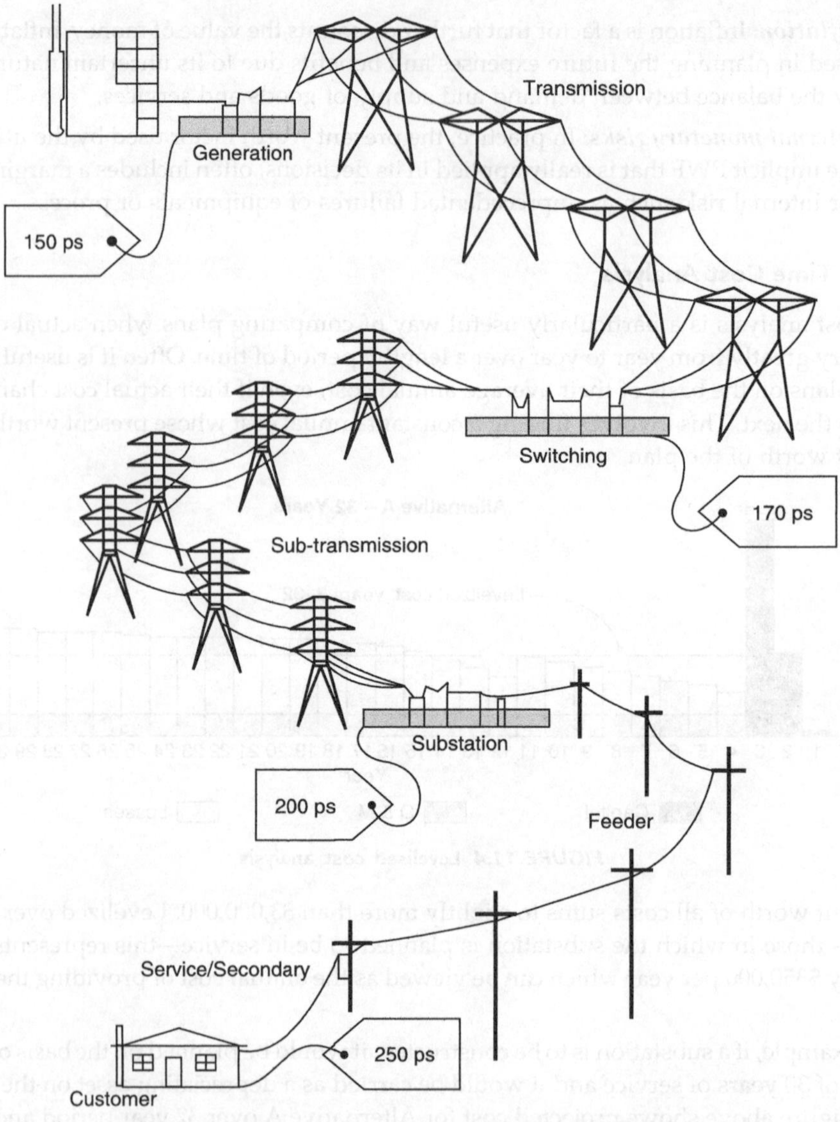

FIGURE 11.5 **Cost variations for different parts of the power system**

twice as far and the cost will double. Thus, from the long term perspective the marginal and incremental costs of delivering electric power are largely unconstrained and linear. However, once a system is in place, its structure is defined, the cost of power delivery varies depending upon the time of delivery, the voltage level of the system through which the power is delivered, the location of delivery, and on any system or equipment contingencies.

The following are reasons for cost variations:

(*a*) ***Temporal variations:*** The cost per kilowatt of power and power delivery varies with season, day of week, and time of the day. The cost of power rises during peak periods because the production cost rises due to the use of expensive generation. Expensive generation is used only when all less expensive generation has already been committed. Delivery cost rises because of increased losses.

(*b*) ***Contingency variations:*** Economy of operation is normally a concern in the design only with respect to normal conditions and no attempt is made to assure economy during emergency. During contingencies, when the equipment is out of service and temporary switching or other measures have been implemented to maintain service, load levels, losses, and loss of life costs are quite high.

(*c*) ***Voltage level of delivery:*** Power is priced less if delivered in bulk at higher voltage, and even lower if consumer owns or pays for the receiving-end power system equipments such as substation, control, relaying and breakers etc.

(*d*) ***Spatial variations:*** Cost to deliver power varies depending on location within a power system. For example, it is difficult to deliver power to some location like the tops of mountains. There may be some consumers staying close to the substation while others staying far away, its expensive to serve the latter. In general the pricing charged for local distribution is based on average cost to serve all customers and is applied without spatial variation.

11.6 ECONOMIC ANALYSIS OF DISTRIBUTION AUTOMATION

The economic, regulatory and business changes facing the utility market are challenging today's executives in many ways. Improving business processes and enabling these process improvements through sound technology provides many opportunities to improve the financial bottom line, reliability and safety, while also improving the ability to deliver the best possible service to customers. It is for these reasons that utilities worldwide are investing in distribution automation technologies that remotely monitor, coordinate and operate distribution components in a real-time mode from a central location. The returns on distribution automation investments are well-documented—systems operate more efficiently, the frequency and duration of outages is reduced and assets are better utilized and maintained. In competitive markets, the resulting cost reductions and reliability improvements help boost competitiveness through the ability to win and maintain customers. But making the right technology choices is much more complex than ever before, as is securing the funding for these IT investments.

To be successful in justifying distribution automation expenditures, utilities need to quantitatively show that the proposed investment will help solve a business problem while supporting corporate strategic objectives and generating a positive return-on-investment. The focus of the analysis must be to develop a strategy that maximizes the system's contribution to the bottom line over an acceptable breakeven period. Only those functions, whose incremental monetary and strategic benefits outweigh the incremental cost to implement the function, should be implemented.

All costing in a power distribution system must be based upon economic evaluation that is consistent over the range all alternatives. Usually, this means that present and future costs must be compared with due consideration to the variations.

11.7 EXAMPLE SYSTEM FOR COST-BENEFIT COMPARISON

An example system of three 11KV feeders from 132 KV Hingna-I substation of Maharashtra State Electricity Board is selected. These feeders being subjected to frequent breakdowns and shutdown, and being neighbouring ones, are the appropriate choice for the distribution automation system. On these feeders, fault event occurrence is more as compared to other feeders of the substation and therefore service interruption to the consumers is quite common. As a result, revenue loss incurred by the Electricity Board due to loss of energy during outage time is considerably high.

The feeders selected for distribution automation are as follows:

(*a*) 11 KV Mankapur feeder from 132 KV Hingna-I Substation.

(*b*) 11 KV Industrial-I feeder from 132-KV Hingna-I Substation.

(*c*) 11 KV Hingna-II feeder from 132 -KV Hingna-I Substation.

11.7.1 Feeder Performance Data

The energy supplied on an average basis to the customer connected to the respective feeders is as follows:

From Mankapur Feeder – 644400 units/month

From Industrial–I Feeder – 552000 units/month

From Hingna–II Feeder – 548400 units/month

Table 11.1 shows the number of breakdown hours and shutdown hours on monthly basis for duration of one year on the three feeders under study. The data has been obtained for Maharashtra State Electricity Board from Hingna–I substation.

TABLE 11.1 Breakdown and shutdown hours for the three feeders of Hingna Substation

Sr. No.	Month	Mankapur Feeder		Industrial-I Feeder		Hingna-II Feeder	
		Breakdown Hrs.	Shutdown Hrs.	Breakdown Hrs.	Shutdown Hrs.	Breakdown Hrs.	Shutdown Hrs.
1	Jun -98	—	15.81	—	0.44	14.48	6.43
2	July -98	—	10.26	—	3.40	—	2.65
3	Aug -98	7.26	8.66	—	6.63	1.05	4.35
4	Sept -98	12.16	13.56	—	8.30	3.48	8.40
5	Oct -98	3.37	1.27	—	2.60	—	4.77
6	Nov -98	5.25	4.45	5.25	0.57	5.25	—
7	Dec -98	0.33	9.22	—	1.52	—	—
8	Jan -99	—	7.36	—	1.45	—	5.10
9	Feb -99	—	16.23	—	6.88	—	—
10	Mar -99	1.45	1.35	—	5.15	—	6.70
11	Apr -99	—	13.55	—	6.34	—	10.03
12	May-99	2.20	2.82	3.0	—	—	1.77

The following are the observations made from the Table 11.1:

1. The occurrence of fault event on Mankapur feeder is highest. This feeder is subjected to more number of breakdown hours as compared to other two feeders under study. Also the shutdown hours on this feeder are large as compared to the other two feeders. So, this feeder is subjected to more number of outage hours and causes greater inconvenience to the customers connected to it and also results in loss of revenue to the electricity board due to the service interruptions.
2. The Industrial-I feeder and the Hingna -II feeder have lesser number of breakdown hours as compared to the Mankapur feeder in a year. However, the three feeders have almost same number of shutdown hours for repair work.

The number of outage hours to the customers on the faulted feeder can be reduced by providing faster fault location and restoration of supply to the healthy sections of the faulted feeder through feeder reconfiguration.

11.7.2 Cost Justification

Table 11.2 gives the installation cost for each component of the system to be installed for distribution automation functions implementation. This information represents typical data.

TABLE 11.2 Installation cost for each component

Sr. No.	Components	Installed limit cost (Rs. In Lakhs)
1	Master Station hardware	43
2	Master Station software license	43
3	Master Station integration labour	43
4	Field RTU communication controller	8.6
5	Role to RTU	3.44
6	Field device communication controller	7.74
7	Remote primary switch package	7.31

Total cost of Distribution automation system = Rs. 1.56 Crores

Calculation of Energy Saved due to Feeder Automation

The outage hours of three feeders are as follows:

1. Mankapur feeder = 132 hrs/year
2. Industrial feeder = 51.53 hrs/year
3. Hingna -II feeder = 75.78 hrs/year

Assuming an opportunity cost of Rs. 4 per unit, revenue earned per year will be around Rs. 2,07,952. It is clear that due to feeder automation, from three feeders only revenue earned per year will be around Rs. 2 Lakhs. If other automation functions such as substation automation are implemented there is an additional 125% reduction in outage hours. So, revenue earned will be around Rs. 4 Lakhs per year.

Though the initial cost of components of Distribution Automation System seems to be quite high, the existing facilities could be used to implement other Distribution Automation functions like, substation automation, automated meter reading, voltage - VAR control , load control etc. , which will provide extra revenue per year in the tune of lakhs of rupees , by way of avoided equipment damage, improved revenue sales, avoidance of failure of capacitor switching , savings in manpower costs, remote metering and improved continuity of service etc. So, the initial high cost for Distribution Automation System is justified by way of additional revenue benefits, which after a few years exceeds the investment cost for Distribution Automation system.

11.8 SUMMARY

Proper methods of costing are required to run any electricity supply undertaking on a sound financial basis. No Utility can survive with disturbing superlatives. Cost is a term inclusive of money, labour, materials, effort, opportunity and other factors. It is usually converted to currency. Basically it consists of capital or fixed costs and operational or variable costs. At the breakeven point the fixed cost equalises with the revenue earned.

Basically there are two methods of economic appraisal: Present worth analysis and Life Time Cost analysis. The cost justification of a Distribution Automation System is done by considering the multiple revenue benefits which it shall provide in a few years.

LONG TYPE QUESTIONS

1. What is cost? Explain its different types.
2. What do O & M costs provide for?
3. Why is return on investment important?
4. What is depreciation?
5. In a distribution system, the revenue (r) varies with time (t) as $r = 2t$, where r is in Rs and t is in years. The total cost (c) varies with time as $c = t + 5$. Plot the above curves along with the fixed cost and find the break even point.
6. Enlist the various reasons for variation of cost in a distribution system.
7. How do the following affect the present worth?
   (*a*) financial risk
   (*b*) inflation
   (*c*) internal monetary risk.

MULTIPLE CHOICE QUESTIONS

1. The amount of money that is required to run an efficient distribution business is called
   (*a*) Capital cost (*b*) Marginal cost
   (*c*) Operational cost (*d*) Embedded cost.
2. Which of the following statements is correct?
   (*a*) Operational costs include the annual cost of taxes, insurance, inspection, etc.
   (*b*) Embedded cost is that portion of the cost which exists in the current land system
   (*c*) The costs of various elements of a power system are always variable
   (*d*) None of the above.
3. The slope of the cost function at the current operating point is called
   (*a*) Embedded cost (*b*) Incremental cost
   (*c*) Marginal cost (*d*) Capital cost.
4. The revenue varies with time as $y = 2.5t$, while the total cost varies with time as $y= 0.5t + 4$, where y is in lakhs and t in years. The fixed cost is
   (*a*) Rs 4 (*b*) Rs. 4 lakhs
   (*c*) Rs. 2 (*d*) Rs. 2 lakhs.
5. In the above question, the cost at the break even point is
   (*a*) Rs. 4 lakhs (*b*) Rs. 5 lakhs
   (*c*) Rs. 6 lakhs (*d*) Rs. 7 lakhs.

6. Temporal variations occur due to

   (*a*) Seasonal changes (*b*) Various contingencies

   (*c*) Changes in location (*d*) Changes in voltage level.

7. If the discount rate is 10%, what is the present worth factor after 2 years?

   (*a*) 0.9 (*b*) 0.82

   (*c*) 0.78 (*d*) 0.64.

8. In the above question, what would be the value of Rs. 100 four years ahead?

   (*a*) Rs. 28.52 (*b*) Rs. 21.76

   (*c*) Rs. 44.44 (*d*) Rs.33.62.

9. If a person has invested Rs. 350 at an interest rate of 7%, after 1 year, the value of his investment would be

   (*a*) Rs. 100 (*b*) Rs. 321

   (*c*) Rs. 800 (*d*) Rs. None of the above.

REFERENCES

1. A.S. Pabla, *Electrical Power distribution,* Tata Mc-Grawhill publication, Fifth Edition, 2004, page 402–424 & 223–261
2. A.S. Pabla, *Electrical Power Systems Planning,* Macmillan India publication, 1998, page 195–250
3. Prof. M. K. Khedkar, *Example System for cost justification, Learning Material for Electrical Power Distribution System.*
4. H. J. Willis, *Electrical Power Systems Planning,* Barnes and Noble publication, Page 135–160
5. Peter Williams & Goran Strbac, IEEE Paper *'Costing and pricing of electricity Distribution services'*, Power Engineering Journal, June 2001, page 125–135
6. Robert Uluski, IEEE paper title *'the benefits and costs of distribution Automation: building a sound business case'*, Electricity Today Journal, January 2004, page 14–16.

Chapter 12

ARTIFICIAL INTELLIGENCE TECHNIQUES APPLIED TO DISTRIBUTED AUTOMATION

12.1 INTRODUCTION

The electrical distribution network is complex in nature and extendable to any extent. So, all the methods almost fail to give exact global solution to the problem. So, we are going approach the Artificial Intelligent Based Systems in order to get exact feasible solution to the existing problem in the distribution network.

In the AI based techniques Artificial Neural Network will provide a solution, if the mathematical configuration is known to us (but electrical distribution networks are complex or zigzag in nature), Fuzzy will provide the range of the solution to the problem, Expert systems provide solution, but due to time varying nature (as time passes, experts are not available, because experts are human beings) it is very difficult to apply. Genetic Algorithm will give the global solution to the problem and it is independent of system network configuration.

Genetic Algorithm can be implemented in many fields- electrical, electronics, aero-space, optimization, intelligent search, etc. In the electrical systems many complex problems can be solved and global solution can be obtained to that problem.

12.2 GENERAL TECHNIQUES DESCRIPTION

The loss minimum re-configuration problem in the open loop radial distribution system is basically one of complex combinatorial optimization, since the normal open sectionalizing switches must be determined appropriately. Therefore, from the limit of the computational burden, such methods as the branch and bound cannot be applied to even a normal scale distribution systems loss minimum problem. Since the problem can be formulated as a mathematical programming model, an approximate mathematical programming solution algorithm is one of the direct type solution methods. The so-called branch-exchange algorithm seems to be the most efficient approximate solution method, and many improved methods have been proposed A heuristic approach is another alternative. Recently, the expert system has been introduced to this area for implementing the heuristic rules. Another method, Simulated Annealing also provides best compare to the other methods but convergence factor is less. Although the above mentioned algorithms can solve the problem with rather less computational burden, the calculation results are only approximate or are only local optimums. To find an approximate global optimal, the Simulated Annealing method is employed. However, the computational burden is too great to get a satisfactory solution.

12.3 GENETIC ALGORITHM AND ITS IMPLEMENTATION

A "Genetic Algorithm" *(GA)* is a search or optimization algorithm based on the mechanics of natural selection and natural genetics. Since GA is suitable to solve combinatorial optimization problems, it can be successfully applied to problems of loss minimum in distribution systems.

It combines the adaptive nature of the natural genetics or the evolution procedures of organs with functional optimizations. By simulating the survival of the fittest among string structures, the optimal string (solution) is searched by randomized information exchange. In every generation, a new set of artificial strings is created using bits and pieces of the fittest of the old ones. It efficiently exploits historical information to speculate on a new search point with expected improved performance.

A set of genes which corresponds to "chromosome" in natural genetics is referred to as "string" in GA. The mechanics of GA are surprisingly simple, involving nothing more complex than copying strings and swapping partial strings. GA starts with a population of strings and thereafter generates successive populations.

So as to evaluate a switch pattern, at first, code them as a finite-length string. A simple code can be generated by considering a string of five 1's and 0's when each of the five switches is represented by a '1' if the switch is on, and a '0' if the switch is off. A random start using successive coin flips might generate the initial population size. After this start, successive populations are generated using the genetic algorithm composed of the following three basic operators:

1. Reproduction
2. Crossover
3. Mutation

These three operators are explained as follows.

1. **Reproduction.** Reproduction is a process in which individual strings are copied according to their objective (fitness) function value (*f*). Copying strings according to their fitness value means that strings with a higher value have a higher probability of contributing one or more offspring's to the next generation. This is an artificial version of natural selection. Let us assume that n strings are generated, and each string has a fitness function value f_i (i = 1, 2... *n*). Then reproduction makes the resulting number of each string proportional to the value calculated by

$$P_i = f_i \Big/ \left(\sum_{i=1}^{n} f_i \right) \tag{1}$$

In *GA*, fitness function must be maximized because the "number of offspring is proportional to its value.

(2) Crossover. After reproduction, crossover may proceed in two steps. First, members of the newly reproduced strings in the mating pool are mated at random. Second, each pair of strings undergoes crossover as follows: an inter position *k* along the string is selected uniformly at random between 1 and (*l* - 1), where, 1 means the length of a string. Two new strings are created by swapping all characters between position (*k* + l) and 1 inclusively. For example, consider strings A1 and A2.

A1 = 0110 | 1

A2 = 1 100 | 0

Suppose we obtain *k* = 4 as indicated by the separator symbol "1". The resulting crossover yields two new strings as shown blow.

A'1 = 01100

A'2 = *l*1001

where A'1 and A'2 are the strings of the new generation.

(3) Mutation. Mutation is the occasional (with small probability) random alteration of the value of a string position. In the binary ding of our black box example, this simply means changing a 1 to a 0 and vice versa. For example, if mutation occurs at the 4th bit of the string, A'1 = 01100.

Then, the 0 – 1 pattern of the string becomes as the following string,

A"1=01110.

Mutation is needed because, even though reproduction and crossover effectively search recombine extent notations, occasionally they become overzealous ad lose some potentially useful genetic and materials. The mutation operator protects against such an irrecoverable loss. It has been suggested that the frequency of the mutation to obtain good results in empirical genetic algorithm studies is the order of one mutation per thousand bit transforms.

General flowchart of GA

Reproduction makes new copied strings for the next generation's mating pool. For all of the strings in mating pool, crossover and mutation are carried out according to the crossover rate and the mutation rate respectively. After these operations, reproduction to the next generation is performed again. These operations are repeated until maximum generation is reached. Some parameters of GA, crossover rate, mutation rate etc., must be determined empirically. The general criteria to determine these parameters are described.

To implement the genetic algorithm to the solution procedure for the loss minimum problem formulated here, firstly, we must determine a string structure and a fitness function.

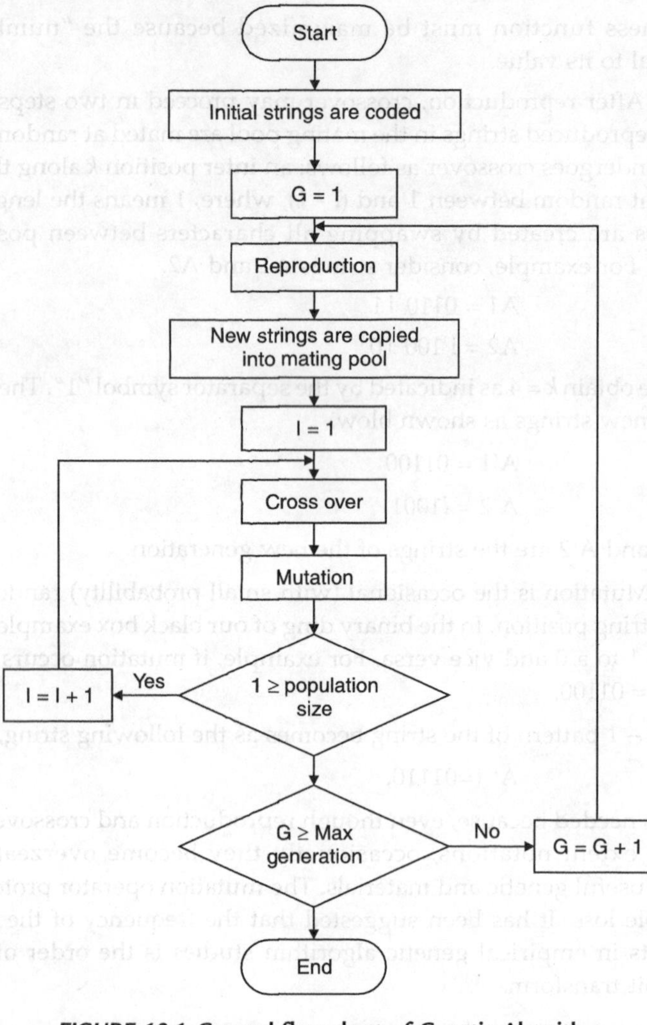

FIGURE 12.1 **General flow chart of Genetic Algorithm**

12.4 HOW GAs DIFFER FROM TRADITIONAL METHODS?

- GAs work with a coding of the parameter set, not the parameter themselves
- GAs search from a population of points, not a single point
- GAs use objective function information, not derivatives or other auxiliary knowledge
- GAs use probabilistic transition rules, not deterministic rules

12.5 STEPS FOLLOWED IN SIMPLE GENETIC ALGORITHM

Following are the steps followed in Simple Genetic Algorithm:

Step 1: Choose a coding to represent problem parameters, a selection operator, a crossover operator, and a mutation operator. Choose population size, n, crossover probability, p_c, and mutation probability, p_m. Initialize a random population of strings of size l. Choose a maximum allowable generation number t_{max}. Set $t = 0$

Step 2: Evaluate each string in the population

Step 3: If $t > t_{max}$ or other termination criteria is satisfied, terminate

Step 4: Perform reproduction on the population

Step 5: Perform crossover on random pairs of strings

Step 6: Perform mutation on every string

Step 7: Evaluate strings in the new population. Set $t = t + 1$ and go to Step 3

The algorithm is straightforward with repeated application of three operators (Steps 4 to 7) to a population of points.

12.6 APPLICATION OF GA TO DA

12.6.1 Chain Table

To accelerate the searching process, chain-tables are used and all feeders can be searched in parallel at the same time. The change in the status of sectionalizing switches *i.e.*, if the bits in a string structure representing a network configuration have been changed, the new network after the change in switch position is easy to be formulated in terms of the bits in a string structure. After the changes in the position of sectionalizing switches it is needed to search for which feeder a given load belongs to. The feeder search technique is employed, as follows:

At first, search begins with the root of a tree. After that the branches linked to the root are searched, their status are set to zero and their end nodes are automatically recorded. These end nodes are used again as a new starting point to search for the other branches and nodes until the searching space is traversed.

Then, the status of the branches is checked for determination of whether a load centre has been shed. If the status of a branch is not zero, it shows that the branch has never been searched and load shedding could occur. Therefore the branch is needed to be added to a corresponding feeder. The reason for this is, when the searching runs into the problem of non-consistence, the branch needs to swap its head and end. Of course, the end node power sink must be simultaneously changed.

To understand this, let us consider a network consisting of following branches {(1, 4, 1), (4, 5, 1), (5, 11, 1), (4, 6, 1), (6, 7, 1)}, where last number of a sub-set indicates status of a branch whose start node is indicated by first number and the end node is indicated by second number. Suppose node '1' is root of the tree. After searching for the tree (feeder), every branch except for (5, 11, 1) can be added into the tree and their status are set to zero. If the status of all the branches in the new network is checked, it is easy to find that the status of the branch (10, 14, 1) is 1, which shows that the branch has never been searched. The branch is not added into the tree and the load centre '11' has been shed. After swapping the head and end of the branch the load centre 11 can be added into the feeder (tree).

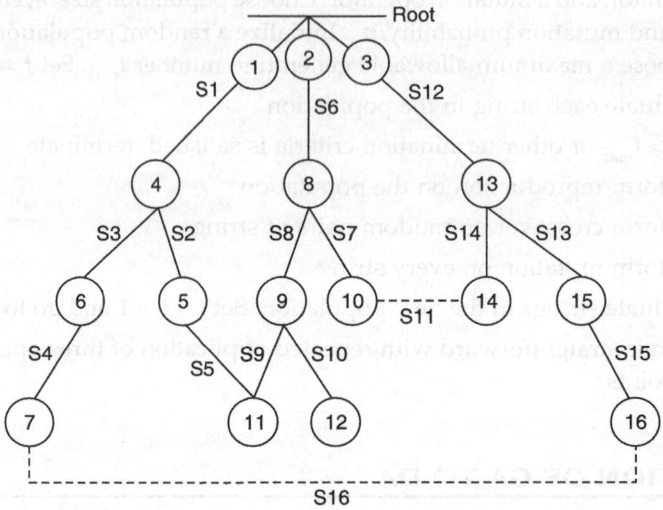

FIGURE 12.2 **Tree of three feeder distribution system**

LONG TYPE QUESTIONS

1. What is an Artificial Intelligence?
2. What are the advantages of AI systems?
3. What are the common AI techniques employed in distribution?
4. How will AI help in improving the performance of the distribution system?
5. What is the Genetic Algorithm? What are the basic operations in a genetic algorithm? Explain them in detail.
6. What is a chain table? How can it be used in a distribution automations system?
7. What are the steps followed in a simple genetic algorithm?

MULTIPLE CHOICE QUESTIONS

1. Perceptron is an example of

   (*a*) ANN   (*b*) Genetic Algorithm

   (*c*) Expert System   (*d*) Fuzzy Logic.

2. The problem with Simulated Annealing is

   (*a*) it is approximate   (*b*) it has a low rate of convergence

   (*c*) it does not yield the desired output   (*d*) none of the above.

3. Consider two strings A = 11011 and B = 00110. After one of the steps of a genetic algorithm, the strings have the values A = 11010 and B = 00111. The step is

   (*a*) Mutation (*b*) Reproduction

   (*c*) Crossover (*d*) None of the above.

4. Consider two strings A = 11011 and B = 00110. After one of the steps of a genetic algorithm, the strings have the values A = 11001 and B = 00101. The step is

   (*a*) mutation (*b*) reproduction

   (*c*) crossover (*d*) none of the above.

5. Consider the following statements

   (*i*) Genetic algorithms work with the coding of the parameter set

   (*ii*) Genetic algorithms search from a population of points

   (*iii*) Genetic algorithms use deterministic values

   (*iv*) Genetic algorithms do not use auxiliary knowledge.

   Of these, the correct statements are

   (*a*) (*i*) and (*ii*) (*b*) (*ii*) and (*iii*)

   (*c*) (*iii*) and (*iv*) (*d*) (*i*), (*ii*) and (*iv*).

REFERENCES

1. M.E.Baran, F.F.Wu *"Network Reconfiguration in Distribution Systems for Loss Reduction and Load Balancing"*, IEEX Trans. PWRD, Vo1.4, N0.2 (1989).
2. K.Nara *"Implementation of Genetic Algorithm for distribution Systems Loss Minimum Re-configuration"*, IEEE Trans.PWRS, Vo1.7, No.3 (1992).
3. Koichi Nara, Atsushi Shiose, Minoru Kitagawa ,Toshihisa Ishihara *"Implementation of Genetic Algorithm for Distribution Systems Loss Minimum Re-configuration"* IEEE Transactions on Power Systems, Vol. 7, No.3, August 1992.
4. *Kyu-Ho Kim, Seok-Ku You, "Optimal VAR Dispatching considering Contingency by Genetic Algorithms"*, Proceeding of the International Conference on Electrical Engineering, Vol. 2, pp.997-1001, August, 1996.
5. S. Sundhararajan, *"Optimal Selection of Capacitors for Radial Distribution Systems using Genetic Algorithm"*, IEEE Trans. on Power Apparatus and Systems, Vol. 9, No. 3, pp.1499–1507, August, 1994.
6. Ahmed R Abdelaziz, Smpeee,Ahmed El-Habachi,*"Hierarchical Genetic Algorithm-Based Switching Device Placement in Distribution Systems"*, IEEE Melecqn 2002, May 7–9, 2002, Cairo, EGYPT.

Chapter 13

URBAN AND RURAL DISTRIBUTION SYSTEMS

13.1 INTRODUCTION

13.1.1 General Concept

Electricity is modern society's most convenient and useful form of energy. Present social infrastructure would be infeasible without it. The increasing per capita consumption of electricity throughout the world reflects the growing standard of living of people.

13.1.2 Distribution of Power

The distribution system is the link between transmission and the consumer service point. It contains:

- Sub-transmission circuit in voltage rating, usually between 33 kV and 220 kV which delivers energy to distribution substation.
- The distribution substation which converts the energy to lower primary system voltage for local distribution and usually improves facilities for voltage regulation of the primary.
- Primary circuit of feeder usually operated in the range of 11 kV to 33kV supplying the load in well-defined geographical area.
- Distribution transformer usually installed on pole or on pads near the consumer sites which transform the primary voltage to the secondary voltage, usually 240/415 volt.
- Secondary circuit at service voltage which carry energy from the distribution transformer along the street etc.
- Service line which delivers the energy from secondary circuit to the consumer premises at declared voltage of 415/240 with an error of 6%.

13.1.3 Basic Distribution System

- Radial
- Primary loop
- Primary selective
- Secondary selective
- Spot network
- Grid network

For details, refer to Chapter 1.

13.2 URBAN DISTRIBUTION

13.2.1 Objectives

A consumer requirement may be summarized as the availability of an adequate and reliable electricity supply at an acceptable cost. Growth in the use of electricity by both industry and commerce and the increasing effects of domestic consumers by their changing attitudes to comfort and labour saving devices introduce additional factors into the already difficult task of load forecasting.

Accurate load forecasting is an essential element in investment appraisal, the long range overall load growth of a city and the necessary super grid reinforcement may be predicted by considering the socio-economic and political climate on the historic load growth curve. However, load growth is rarely uniform throughout a city, and small area land use analysis provides a sound basis for medium term strategic planning. For detailed short and medium-term planning and for system optimization it is necessary to know the exact location of loads, and the use of a computer program to estimate maximum demands from the customer billing records is a vital supplement to accurate transformer loading data.

The major investment costs will be largely determined by the supply arrangements and security standard adopted by the supply company. The distribution system, providing and reinforcing assets between major supply points and the consumer terminals relates to the development of the distribution system.

13.2.2 Electrical Consideration Factor for Cities

Characteristic Factors

A large number of cities in the world have areas of approximately 100 km^2 with the density of buildings and electrical load increasing rapidly towards the centre. Frequently, such cities are also divided into two or more parts by rivers or waterways which present problems to designers in obtaining ideal routes for underground cables.

A typical city undergoing rapid development would have load densities in the outer areas of the order of 5 MVA/km^2 or less rising to 60 MVA/km^2 or more in the centre, with a total demand of perhaps 1250 MVA within about 15 years time.

The supply zone for each primary substation is necessarily much larger for the outer and middle bands of a city than for the dense central areas, and more substations are required in the central area than the outskirts. Typically with primary substations having an installed capacity of 2x60 MVA and loaded to a maximum of 114 MVA, five substations would be required on the outskirts, four in a middle band and perhaps five/six near the centre.

(*a*) **System Development and Security of Supply.** The types of generating plants installed, voltage ratings and other system parameters are directly linked to the historical development of the major activity centers within countries and the technological advances in engineering. There is seldom an opportunity to forget what is there already and start a new, as capital investment must be carefully balanced against anticipated revenue from the installation of new or modification to existing assets. Security of supply at a reasonable price is an important aspect of the supply to a major city. By careful design the security of supply can be achieved as a byproduct of the overall design with little direct attributable cost.

TABLE 13.1 Recommended classes of supply and security levels in UK

Class of supply	Range of group demand, MW	Minimum number of supply circuits normally available	Demand to be met immediately after First circuit outage	Demand to be met immediately after Second circuit outage	Target time for the restoration of full group demand
1*a*	up to 1	1	nil	—	for first circuit outage repair time
1*b*	Over 1 to 8	1	nil	—	two hours
1*c*	Over 8 to 24	1	nil	—	15 min for second circuit outage
2*a*	Over 24 to 80	2	group demand	nil	repair time for one circuit
2*b*	Over 80 to 300	2	group demand	nil	repair time for one circuit full group demand, but one-third group demand in two hours
3	Over 300	3	group demand	all consumers at 75% group demand	repair time for one circuit

Most countries work within a nationally agreed framework which sets out agreed standards for the security of supplies to various classes of consumers. Table 13.1 shows the table of recommendations for the United Kingdom. These standards provide a firm basis for affording supplies to a modern city but, with forethought, improved designs can be achieved at a very moderate premium over an absolutely basic system.

(*b*) **Consumer Categories.** In UK, majority of the electricity consumers are domestic industries which take about 37% of the total energy supplied. In India, electricity consumption of industry is 37%, commercial—13%, agriculture—26% and domestic—24%. In rapidly expanding distribution systems the maximum demand frequently occurs for a portion of the day and the overall utilization of the assets and the economics of the system can be significantly improved by the control of specific load categories such as heating and chilling to fill the daily load-curve valleys. The central control of such loads and associated tariffs can be achieved by SCADA systems.

(*c*) **Standardization.** With a large city which has a high rate of electrical load growth, it is vital that there is an overall strategic plan and that supplies to individual areas or consumers be provided quickly, cheaply and in such a manner as to advance the overall plan. For the above reasons, the design of each circuit element should be standardized so that minimal detailed design work is necessary for each new job, thereby reducing the costs and the technical resources required. A standardized approach makes the application of automation to a system practical at an economic cost and without the need for constant change by highly skilled staff.

(*d*) **Automation.** To achieve a consistently high level of security of supply and to reduce the dependence on technical staff to restore supplies, the maximum use should be made of system automation and control computers. By interlinking substations and grid points with fiber-optic cables laid with extra-high-voltage (EHV) and high-voltage (HV) circuits, the whole city system can be controlled by a minimum of staff from one control centre. An unmanned backup control centre can be provided for use in the event of a catastrophic failure at the control centre. This approach is widely adopted both in the UK and in the Western world as a whole.

(*e*) **Maintenance.** A modern system should be designed to require minimum of maintenance; switchgear should be reduced to an absolute minimum and be of SF_6, or vacuum type, and transformers should preferably be sealed units. Equipment should be well insulated and placed in a dust-excluding housing. Protection equipment should be kept to the minimum number of locations and pilot wires avoided .

13.2.3 Primary Systems

Bulk in-feeds of power to a major conurbation, whether from local generation or remote sources, are only likely to be available at a limited number of points on the perimeter of the conurbation. To interlink these in feed points and to provide the means of transmitting large amounts of power closer to the points at which it is required can best be achieved by the establishment of an EHV dual circuit ring around the perimeter of the area. Typical operating voltages for such a ring would be in the range 220–500 kV dependent on the existence and suitability of present assets. A typical 4x400mm Aluminum Conductor Steel-reinforced (ACSR) installation operating at 500 kV would have a capacity of about 3500 MVA.

At strategic points, depending on load requirements, grid substations (super grids) would be established around the ring to enable power requirements to be drawn from the system transformed down to a more manageable voltage (100–150 kV) for transmission towards the city centre.

Typical switchgear for both voltage-level applications could be either air blast or sulphur hexafluoride (SF_6). Typical switchgear layouts at a super grid could include tees and mesh corner connections.

Choice of numbers in EHV/HV Transformers and Circuits

To meet the connected load for a first circuit outage the usual approach is to run banks of two or three transformers from two EHV lines; this results in a maximum utilization of 65 and 86%, respectively.

Where the availability of land for substations at an economic price becomes a problem, HV load-transfer facilities can be utilized to obtain overall transformer and circuit utilizations in the range 95–100% resulting in a saving of 16–50% in the number of substations required for a given load. Furthermore, where the use of overhead lines for EHV circuits is inappropriate and underground circuits are necessary, particular emphasis is placed on achieving a high utilization of these very expensive assets, the cost of which often swamps that of the substation building and equipment.

In summary, substations with two and three transformer installations are the norms for rural and light urban areas where overhead EHV circuits are permissible. For city centers where land for substations is unavailable or expensive and EHV circuits must be underground, the high utilization achievable with HV load transfer facilities is essential.

Arrangements with one, two, three and four transformers are in use in the UK each of which has advantages and disadvantages:

1. Transformer substations require a very large number of sites, an equal number of EHV cable routes which may be a problem in narrow city streets and a very large investment in HV load-transfer facilities which are rarely used.
2. Transformer substations are optimum for most city situations in that the HV transfer facilities can usually be achieved without great additional expenditure and can be deferred until the load approaches the designed firm capacity (78 MVA) of the substation.
3. Transformer substations are preferable to two where the load density is particularly high (over 20 MVA/km^2) over a considerable area, to reduce the number of sites required in close proximity. Their use in lighter loaded areas is to be avoided in view of the number and length of HV feeders required to match the large supply zone.
4. Transformer substations are not very practical with dual-winding 60/30/30 MVA owing to the complexity of the sequence switching required, the large sites and difficulties in restricting fault levels. Such installations would only be appropriate in extremely high load density areas, *i.e.,* consistently above $100MVA/km^2$.

13.2.4 High-voltage System

The power drawn from the EHV system and transformed down to HV level (100–150 kV) is transmitted to the load areas over two circuits. These two circuits have the capability to supply one or two substations, located at or near load centers, at which the power requirements would be transformed down to lower levels (10–20 kV) for distribution into the local HV/LV networks. This enables an economic selection to be made in grading the size of HV cables used in the super grid. Typically it is envisaged that from the super grid to the first substation (630 mm^2) cable would be used whereas between the first and second substations the cable size could be reduced to 185 mm^2.

Connections from the EHV ring are arranged to optimize system security by using mesh corners, and alternate connections from each circuit.

A typical new HV switchgear installation would utilize vacuum or SF_6 equipment to control both the HV side of the 60 MVA transformers and the outgoing distributors to the local network. The advantages of the non-oil type equipment, with respect to minimum maintenance, reduced space requirement, and the removal of the need for fire protection within the switchgear areas are particularly important in congested areas where main substations have to be located within the other buildings.

(*a*) **EHV/HV Substation Loading.** A 2 × 60 MVA substation based on the concepts outlined should be capable of supplying at least 78 MVA under outage conditions of a single transformer, which gives a utilization of 65%. By using suitable bus bar configurations and projecting the design of HV and LV distribution networks associated with the substation it is possible to substantially improve this level of utilization. The normal loading of a 2 × 60 MVA substation could be increased to 114 MVA, giving a transformer utilization of about 95%. To cater for this increase in utilization, automatic transfer capabilities using the associated HV network is established between adjacent EHV/HV substations so that the load in excess of 78 MVA can be transferred to adjacent substations.

On the occasion of a double-circuit outage, the load can still be supported provided the system is operating at about two-thirds of total load.

(*b*) **EHV/HV Substation Location.** To avoid the costs associated with procurement of land for new substations, it has become common practice to provide purpose-built EHV/HV substations either partially or totally underground or within the basement areas of existing buildings.

Modern developments in both HV switchgear design and transformer materials have over the years enabled the space requirement for EHV/HV substations to be progressively reduced. A modern EHV/HV substation comprising of two 60/30/30 MVA 132/11/11 kV transformers, associated 132 kV open-type disconnectors , 47 panel HV switchboard, and all associated auxiliary equipment can now be accommodated within an area measuring 50 m × 25 m × 10 m high (12500 m^3). In the event that such a substation is accommodated totally underground, ventilation requirements to permit the passage of air to and from the transformers can be accommodated with inlet and outlet ventilation of areas 30 m^2 each. Alternatively when the substation is within another building, developments now permit the heat produced by the transformers to be used to supplement the heating requirements of the building.

The small space requirements for modern substation design opens up the possibility of redeveloping existing sites and increasing the capacity by 3–5 times, the load of the existing substation being temporarily transferred to a transportable package substation which can be located on nearby open land such as parks or car parks. When the redevelopment is complete the package can be moved on to the next substation where similar extensions can be planned.

13.2.5 Secondary Distribution System

The secondary distribution voltage should be in the range 10-20 kV to optimize the cost per MVA of power distributed, the actual choice being dependent on load density patterns. HV cables leaving primary substations are organized into groups of four radial feeders with each group having a nominal maximum load of 18 MVA. By the provision of suitable HV switchgear each group is arranged to be firm for the loss of a single circuit following manual or automatic switching.

The general network arrangement is ***'loop-tee-loop'*** involving the use of ring-main switches at the looped substations and some teed substations where the feeder is to continue on to another substation. At the primary substation, protection to the cable and transformers should be provided by a microprocessor relay which provides over current and earth-fault protection for the feeder circuits and transformers. The device functions by detecting the step change in susceptance. Such an arrangement enormously simplifies routine maintenance. The use of a faulted section indicator system such as *'Visit race'* or *'Reyrace'* enables the location of HV faults to be rapidly undertaken.

13.2.6 High Voltage /Low Voltage Substations

The LV secondary voltage should conform to international standards and lie within the range 400 V ± 10%. The UK standard is 415V ± 6 % but the actual voltage chosen should, if possible, match that already in widespread use. With a 400V three-phase LV supply the economic maximum supply distance is around 400 m. Assuming that the city has a grid of streets this gives a supply zone having a radius of 400 * 1.42 = 280m which is an area of approximately 0.25 km^2. Although locations of substations can never be ideal, it is clear that a minimum of five substations are required per squared kilo metre. With a minimum load density assumed of 5 MVA/km^2 a choice of 800 or 1000 KVA transformers would be most suitable.

13.2.7 LV Systems

The object of an LV system is to supply the growing load at both the smallest initial and the lowest incremental cost. The system for the city must be capable of continuing development without repeated re-excavations and overlaying in the same areas. The well designed system will also provide a measure of security of supply to provide for retaining supplies while one distribution transformer is maintained or replaced after a fault.

To match the capacity of the distribution transformer and meet all the loads in a street except those large enough to warrant a transformer themselves, a spine circuit of at least 400A/300 KVA rating should be laid along at least one revenue-earning frontage of each major street. The distribution should run initially directly between substations. 240 mm^2 aluminum cross-linked polyethylene (Al XLPE) or waveform combined neutral and earth (CNE) cable laid direct at a depth of 450–600mm would be an excellent choice being low cost but having a high thermal overload rating, easy to joint and lay in a congested city area. Such cable is widely employed in the UK electricity supply industry, particularly in London.

It may be seen that, once the full mesh of distributors has been completed, it is generally unnecessary to add further LV cables to match the growing loading. Services to individual buildings should be provided by cross-linked polyethylene (XLPE) or polyvinyl-chloride (PVC) cables of suitable size 240/185/95/35 mm^2 to insulated cased secure fused terminations, preferably located in recesses on the outside of the premises, connection to the LV distribution being by cold pour resin service joints.

Crossroad services will be necessary where initially only one cable is installed in a street. For areas of cities which have consumers located in a multiplicity of narrow alleys rendering excavation for cable lying impracticable, the introduction of aerial bundled conductors attached at a suitable position directly to the property frontages is recommended.

The system in its basic form consists of two or four XLPE insulated conductors laid up in a helix to form a bundle of the conductors, which may be sized between 25 and 95 mm^2 and are of stranded hard-drawn aluminum which is partially compacted. Additional small cores may be added to the main bundle for use as street-lighting switch wires or for other control functions.

There are many advantages resulting from the choice of this system when compared with conventional open two or four-wire bare conductor installations and these include increased safety, ease of operation, quicker installation, and improved reliability of supply, lower overall costs and it is aesthetically more acceptable.

13.2.8 Modernization of Distribution System

The advent of pole mounting reclosers; Sectionalisers, switch capacitor bank and auto voltage boosters made it possible to modernize the exiting system, using

Auto reclosers

These are low rupturing capacity circuit breakers which open out on fault sensed by inverses time delay with instantaneous provision and reclose at preset number of times at preset time interval and get locked in open position, if fault persists.

Sectionalisers

These do not interrupt fault, but the line can be sectionalized into various sections. By using them the faulty section gets isolated. These operate by sensing whether the line is ON or OFF and various type of fault detecting relays and control boxes make these adoptable for radial, loop and ring feeder etc.

Switched capacitor bank

These are switched on and off automatically to put into service when needed only and avoid over compensation during light load period. The control system operates by sensing current (load).

Automatic voltage boosters

Essentially an autotransformer with a shunt and series winding is provided with taps on the series winding and the taps are automatically selected based on voltage level. AVBs can boost up to +10% in step of 2.5%. AVBs can also provide voltage variation for 10% in 32 steps of 0.625% each.

13.2.9 Information Technology Initiatives in Electrical Distribution

- Consumer analysis tool (CAT).
- Book consolidation module (BCM)
- Circle performance management and reporting system (CPMRS)
- Monitoring and tracking system(MATS)
- Transformer information management system (TIMS)
- Micro controllers

- Metering, billing and collection (MBC)
- GIS (Geographical Information System) and GPS (Geographical Positioning Systems)
- Distribution automation

(*a*) **Consumer Analysis Tool (CAT).** The objective of Consumer Analysis Tool (CAT) focus on generating prioritized and focused consumer information relating to all business aspects at summary and detailed level for LT categories. The objective of the tool is to analyse customer information to identify trends/anomalies in metering, billing and collection for each customer and provide detailed information on the same to enable further action.

Benefits of CAT:

- Aids in analysis metering, billing and collection of data.
- Generates priorities and focused report for action initiation.
- Aids to reduce the response time for report generation and action initiation.
- Acts as an effective monitoring tool, helping management in exercising control.

(*b*) **Books Consolidation Module (BCM).** The objective of the BCM is to reduce the time taken and the manual intervention involved in the consolidation of accounts from the circle level upwards. The module uses combinations of an accounting package as well as Microsoft excel to achieve this.

Benefits of BCM are as follows:

- BCM has helped to reduce the processing time required for consolidation from nearly 60 days to around 15 days
- The management analyse the performance against budgets and maintains control over budget.

(*c*) **Circle Performance Management and Reporting System (CPMRS).** The Circle Performance Management and Reporting System is designed as an executive information system to assist top management to monitor the performance at each circle. The system facilitates the performance management process by measuring and reporting performance using which the companies can review, provide feedback to offices concerned and plan action.

Benefits of CPMRS are:

- Operation statistics
- Distribution network monitoring etc.

(*d*) **Monitoring and Tracking System (MATS).** The MATS has been designed as a tool to assist the process of monitoring and tracking of the progress of all cases of irregularities, like theft, pilferage, malpractice, back-billing etc.

Benefits of MATS are:

- Greater transparency and accountability.
- Ensures adherence to timelines.
- Ensures automated document flows.
- Prioritization of inspections and follow-up to optimize resource utilization.

(*e*) **Transformer Information Management System (TIMS).** The objective of TIMS is to ensure effective tracking of distribution transformer as well as associated structure thought—out their life cycle and analyzing.

Benefits of TIMS are:

- Location with maximum transformer failures.
- Identification of transformer and structure which show frequent failures.
- Time required for repairing transformer at each repairing location.

(*f*) **Micro controllers.** Micro controllers are provided at each of the 33 kV substations to achieve the following:

- Ensuring the number of hours of power supply including implementation of restriction and control measures when in force.
- Automatic tripping of supply, if load exceeds a preset value during specific hours to avoid over loading of equipment.
- Monitoring of uniform distribution of load as per schedules.
- Reading of all event and data parameters is provided to generate reports data feeder-wise, hours of supply and complete data required for conducing load survey.

(*g*) **Metering, Billing and Collection (MBC)**

- *Common Billing System (CBS).* Common software for both HT and LT billing, revenue collection, new customer accounting ,change of ownership of service, dismantlement and termination of service etc., is developed so that the entire MBC function is done at a common platform.
- *Spot Billing System (SBC).* Spot billing is done trough hand held computers which are fast, versatile, intelligent, affordable, portable, and user programmable.
- *Remote Metering.* The meters at selected high value of service are read remotely and consumption pattern is monitored. Meters tampered or not recoding properly are instantly detected. Cases of pilferage are also detected as meters have tampered proof data.

13.2.10 GIS and GPS

Geographical Information System (GIS)

GIS is a computer based tool used to integrate the data of system network. GIS is an organized collection of computer hard ware, software, geographical data and personnel which can efficiently capture, store, update, manipulate, analyse and display all forms of geographically reference information.

GIS when integrated with real time SCADA helps in efficient handling of customer inquires, fault management, preventive maintenance, network optimization, energy audit, improved revenue, accurate power flow results and short circuit analysis etc.

Global Positioning System (GPS)

GPS is a satellite-based navigation system. The satellites are oriented in such a way that from any place on the earth, at any time, information is available for navigational purposes.

GPS survey is done to capture latitude and longitude of network facilities like substation, transformer, HT and LT poles etc. and transfer the data directly to the GIS system for analysis.

Applications of GIS and GPS:

- Network analysis, optimization and planning
- Power supply management
- Energy accounting and revenue protection
- Transmission studies
- Trouble call management
- Automatic meter reading, estimation.

13.3 RURAL DISTRIBUTION SYSTEM

About 35% of the total power generated in India is consumed by over 70 per cent of the country's rural population. Concerted endeavourer must be made to raise this consumption to 50 per cent of the total generation, which can serve as a good index of India's rural prosperity.

13.3.1 Objectives

- To improve the economical status of the rural population by increasing the productivity of human and animal labour
- To promote rural welfare by providing an environment equal in comfort and convenience to that enjoyed in urban areas. Thus electrification is an important factor in preventing the drift of the rural population to urban areas, where higher wages and more comforts are available.

Reliable supply to rural areas will reduce slum areas in cities.

13.3.2 Reliability

Reliability and quality of supply refer not only to the number of complete outages the system may suffer but also refer to any deviation from the design limit of voltage , frequency, waveform ,current carrying capacity, voltage flicker and system impedance up to the point of utilization .

Any of these may affect industrial or commercial consumer, but most relevant ones as regards rural consumer are voltage deviation and outages. However, some industrial consumers who are occasionally supplied from rural supply system may experience serious outage hazards.

13.3.3 Faults and Protection in Rural Distribution System

- *Lightning*: Lightning causes phase to earth or phase to phase insulation flashover, followed by switchgear or drop out fuse or fuse operation and sometimes pole or cross arm fires. It can also cause failure in distribution transformer and occasionally, direct strike can destroy poles and break conductors. Lightning over voltage protection is important and the installation of rod gaps or lightning arresters on HT side of pole –mounted transformer is desirable.

- *Animal and bird life*: These are generally causes of nuisance to transformer poles where phase to phase and phase to earth clearance are smaller than on normal line poles. Bird guard on transformer poles and insulation using polythene tubing or bushing caps, where clearances are less, can prevent most of these problems.

Other remedies are as follows:

1. To increase inter phase clearance or use 'V' or 'U' type cross arms for lines near bodies of waters such as ponds, rivers and canals.
2. To avoid pole top designs which provide vertical clearance on such places, and animal burial grounds nearby.
3. To mount suitable PVC sleeves on the affected portion of conductors.

- *Mechanical failures*:

These can occur due to:

1. Defective conductor.
2. Felling of trees or tree branches on line during stormy days. Cross-arm and pole can fail for similar reason.

- *Man made fault*:

Vandalism, such as stoning, shooting at insulator, throwing of chains/wire on conductor etc., can cause troubles resulting in permanent faults. Broken insulator may not be traced for weeks and cause pole or cross arm fires in the case of wooden structure.

Protection

- *Sectionalizing devices*: These devices at suitable locations functionally improve the reliability of services by isolating faults to a portion of a circuit and by reducing the fault locating time. These are of the type of auto-reclosers, fuse cutouts, circuit breakers and load switches.
- *Power transformer operation*: Generally frequency of faults on rural feeder is more due to long overhead lines and their vulnerability to natural hazards. At main or grid substation in rural areas having two or more transformers, the transformers should not be run in parallel, so as to keep the fault level low. This will facilitate the reduction of the faults strain on breakers, cable, lines etc., and will enhance their lives and improve the continuity of supply.
- *Network protection*: Rural networks, being lengthy and scattered need proper protection. Proper fault location, proper sectionalizing for speedy removal of faults and continuity of supply should be ensured. In most of the electricity boards, primary protection is provided for the rural distribution transformer, with isolating switches in the HT fuse. Secondary protection is provided with LT fuse cut-out.

13.3.4 Improvement of Existing Distribution System

Any design for improving the system should at least cover the whole of sub-transmission and distribution in the given area and therefore comprises of relevant levels of voltage such as 132, 66 and 11 kV.

Poor power factor, high distribution losses and poor voltage conditions in rural system are generally because of following factors:

1. Long and overloaded 11kV and sub-transmission lines.
2. Low power factor of agriculture tube well pumps at consumer installation.
3. Absence of shunt compensation in the sub transmission and distribution system.
4. Location of distribution transformers center away from the load centers.

Remedial measures

1. Re-routing the main link line of the feeder by shortest route.
2. If re-routing is not possible, then the feeder should be studied for bifurcation at suitable points with a view to transferring loads to lightly loaded feeders.
3. Provision of shunt compensation on 11kV feeders can be made.
4. Whenever the distribution transformer is overloaded and additional loads are anticipated, the exiting transformer can be replaced by a higher voltage transformer. Alternately, a new transformer center may be provided to cater for excess loads.
5. When power transformer in the HT sub-station is fully loaded, the augmentation of the transformer capacity can be carried out either by installing an additional transformer or by changing the exiting transformer with a higher size unit.
6. Provision of new 11 kV LT feeder and distribution transformer can be made.
7. Power factor at the load point can be improved considerably by installing LT capacitor at the consumer premises and at the distribution transformers.

13.3.5 Single Wire Earths Return System (SWER)

For remote rural areas with sparse population, the SWER system has been found to be economical. Here the earth is used as the return conductor and one wire as the phase conductor. This system requires grounding rod deep into the soil at each consumer site in order to establish sufficient low impedance to ground. This system is used in India (Madhya Pradesh, Chhattisgarh, etc).

Rural electrification with SWER is an inexpensive method of extending the electricity grid to rural areas. SWER will reduce electrification costs to rural communities in developed and developing countries. Rural electrification using SWER will directly and indirectly help to increased profits of utilities.

In an SWER system, there is only one wire instead of;

(*a*) Two wires used in Europe for single phase distribution or;

(*b*) Three wires used in Europe for three phase distribution or;

(*c*) One phase wire and one neutral wire used in the North American single-phase distribution system.

To summarize, the *advantages* of SWER are:

1. Economical to the extent of 30–40 per cent over conventional three-phase system under favorable circumstances.
2. Better voltage regulation which is the main problem in long conventional feeder.
3. Improved power factor as compared to the conventional system, being 0.95, whereas it varies from 0.7 to 0.75 for the conventional system.
4. The outages on the SWER system are comparatively lesser than on the conventional system due to the less number of components per span.

The main *limitations* in adopting the system are:

1. Maximum ground return current permitted is 7–8 A.
2. Maximum power limit of distribution of this system is limited up to 3–3.5 KVA/km of the line.
3. Particular configuration and the maintenance of earthing electrodes is an important point while adopting this system.
4. The SWER system without the isolating transformer might create extraordinary interference with telecommunication circuits.

13.3.6 Locating Faults

In accordance with the present practice in the rural distribution system, the main 11 kV feeders are protected by a circuit breaker with over current and earth fault relays.

In case of sustained faults, the circuit breaker does not close and the field staff is required to patrol the main feeder as well as the sectional lines to locate and isolate the faulty section before supply is restored.

With a view to eliminate the possible delay in locating and isolating the faulty section, an 11 kV overhead fault locating equipment can be installed on the line at tee-off points .The usual faults locating equipment installed on HV or EHV lines are the pulse reflection locator and distance relay with distance measurement unit. The unit for rural long lines must be cheaper and operationally viable.

13.3.7 Auto-Reclosers

The reclosers can be used:

- As the primary feeder protection device at the rural sub-station.
- On major tee-off lines to protect the main feeder from interruption due to faults on the tee-off lines.
- For sectionalizing lengthy feeders in order to prevent outages on the entire feeder, when permanent faults occur near the end of the feeder.

13.3.8 Future of Rural Distribution Systems

- Adoption of reduced insulation level
- Segregation of rural supplies into:
  1. Agriculture supply
  2. Rural industries
  3. Rural lighting
- Power voltage transformer
- Decentralized generation and distribution.

LONG TYPE QUESTIONS

1. What are the components involved in distribution of power?
2. What are the different types of distribution system?
3. Explain Urban Distribution.
4. Where should a EHV/HV substation be located?
5. Explain role of sectionalisers and switched capacitor banks.
6. How do we overcome faults in rural distribution system?
7. How does a Single Wire Earth Return System reduce the cost of distribution?
   What is auto reclosure?

MULTIPLE CHOICE QUESTIONS

1. Primary circuit of feeder operates in the range of

   (*a*) 10 kV–15 kV (*b*) 11 kV–33 kV

   (*c*) 66 kV–77 kV (*d*) 440 kV–1200 kV.

2. Loop Tee Loop network arrangement uses

   (*a*) Ring Main switches (*b*) Oval Switches

   (*c*) Circular Switches (*d*) Capacitor Banks.

REFERENCES

1. Graham Dennis, "*Electricity Distribution System in Urban areas*" IEE Power Engineering Journal November 1998. Page(s) 332–336.
2. J.P. Green, S.A. Smith, G. Strbac, "*Evaluation of Electricity Distribution System Design Strategies*" IEE Pro-Gener. Transm. Distrib. Vol. 146 No. 1 January 1999. Page(s) 53–60.
3. A.S. Pabla " *Electrical Power Distribution*", TMH, 5th Edition, 2004.
4. M.V.S. Birinchi. "*Modernization of Power Distribution*", NPTI, First Edition: March 2004.

Chapter 14

Optimal Switching Device Placement in Radial Distribution System

14.1 INTRODUCTION

Any low voltage radial distribution network is vulnerable to faults. In the event of a fault anywhere in the network the circuit breaker trips, causing outage of the total system network ahead of switching device. Thus, sectionalizing switches are excellent solutions to improve power quality, VAR control, and even synchronous closing applications.

14.2 SECTIONALIZING SWITCHES

These are no load switching devices used in association with auto re-closing circuit breakers for overhead distribution feeders. They provide isolation for a certain section of overhead feeder. Sectionalizers with auto re-closer are used universally for rural distribution lines and radial overhead distribution feeders. Automatic sectionalizers are no-load switching devices, which are provided with over current relays, earth fault relays and counters. The counter counts the number of operations of the back up circuit breaker or auto re-closer. After a predetermined number of back up circuit breaker/auto re-close, if the circuit breaker/auto re-closer has not cleared the line fault, the main circuit breaker/auto re-closer opens and thereafter the automatic sectionalizers open under no current condition.

The automatic sectionalizers are set to open after predetermined number of tripping operations of back-up auto re-closer. This number of predetermined operations of backup breaker is pre set for in series after an auto re-closer.

The operating modes of automatic sectionalizers are summarized as under:

1. If the fault on overhead line is temporary and is cleared while auto re-closer is open, the connector of sectionalizer is rests automatically to its normal position after re-closing of the circuit by auto re-closer.

2. If the fault on overhead line continues when auto re-closer is closed, the fault current counter in the sectionalizer will prepare for counting of the next opening of the auto re-closer.
3. If the auto re-closer is set to lock open after fourth opening operation, the automatic sectionalizer will open during open circuit time the third tripping of the auto re-closer.

14.2.1 The Need

In a given feeder network (in the absence of sectionalizing switches), whenever there is a fault anywhere in the network, the main circuit breaker would trip and all the areas to which the distribution network is supplying power would go into dark. This is not desirable as the electricity board will have to bear huge losses due to the non-supply of power even to healthy areas.

Now let us consider the presence of sectionalizing switches at various locations in the feeder network. Whenever there is a fault in any section of the feeder, sectionalizing switch nearest to the fault location opens first. The service continuity of the main feeder and other branches is maintained. The faulty branch sectionalizer is locked open. Thus the outage area *i.e.*, the area in dark is significantly reduced and hence revenues of the Electricity Boards are not completely cut out.

14.2.2 Types

Sectionalizing switches used for 11kV, 415V distribution networks are basically of three types:

1. Air Break Switches
2. Load Break Switches
3. Molded case circuit breakers (MCCB)

For High voltage distribution systems vacuum break line sectionalizers are used which are more expensive.

The above mentioned switches can be operated from a remote location, ideal for a SCADA system. In distribution systems there are certain proven methodologies and innovative options to determine the right reliability scheme for the system. The scheme should be such that it helps to improve reliability and profit margins by combining both, prevention and outage response strategies.

14.2.3 Benefits

Benefits obtained from installation of sectionalizing switches are:

1. Immediate isolation with indication of the faulty section.
2. Immediate restoration of supply to healthy section
3. No down time for transient faults
4. Improvement of revenue due to lesser outage of section and low down time.
5. Requires less manpower in the system
6. Better reliability of power supply.

14.2.4 Additional Circuitry

As mentioned above sectionalizing switches cannot be installed anywhere in the system because of cost constraints. In addition, the use of sectionalizing switches in SCADA requires additional circuitry like optical fibres and communication lines for proper protection with the RTUs. Hence the sectionalizing switches should be located at only those points, wherever the cost and location is optimum, *i.e.*, maximizing the profits and minimizing the outage area *i.e.*, the area in dark.

The proposed algorithm

The algorithm is developed which is based on Bellman's Optimality Principle. It proposes a method by which optimum number and location of sectionalizing switches can be determined for radial low voltage distribution networks.

Bellman's optimality principle's major issue is cost consideration. Due to high probability of occurrence of faults and rising energy costs, emphasis is given on reducing the outages occurring in the system by placing more sectionalizing switches in distribution system. This number is restricted due to high cost of device and low revenue earned in sub-urban networks. Thus the Bellman's Optimality Principle states that "An optimal policy has the property that whatever the initial state and initial decision, the remaining decision must constitute an optimal policy with regard to the state resulting from last decision".

The above principle in terms of distribution system is redefined as "Optimizing the location and the number of sectionalizing switches such that the resulting benefits, in terms of reduction in cost of outage, justify the capital investment for their installation".

The basic idea is that the cost of outage of the complete network in absence of switching device is calculated. The cost of outage of complete network in the presence of switching device place at position calculated. The difference in the outage cost obtained gives the idea about the loss/profit that can be incurred. Hence we are trying to maximize the benefit obtained by placement of device in various locations and minimizing the cost of disruption.

The proposed formula that calculates the cost of outage is given as

$$c_d = \sum_{j=1}^{N_b} C_{kwh}\lambda_j L_j \left[\sum_{i=1}^{N_{Loc}} P_i t_{loc} + \sum_{i=1}^{N_{rep}} P_i t_{rep} \right]$$

where N_b = Number of branches

λ_j = Number of faults for every 0.2 km of branches

L_j = Branch length km

C_{Kwh} = Cost of non-supplied energy

N_{loc} = Nodes isolated during fault location

N_{rep} = Nodes isolated during fault repair

P_i = Node power

T_{loc} = Time for fault location

T_{rep} = Time fault repair.

Using principle of optimality, it is merely now to maximize the benefit obtained by the placement of device in various location and to minimize the cost of disruption. The flow chart for proposed algorithm to optimize the number and location of switches in shown in Fig. 14.1.

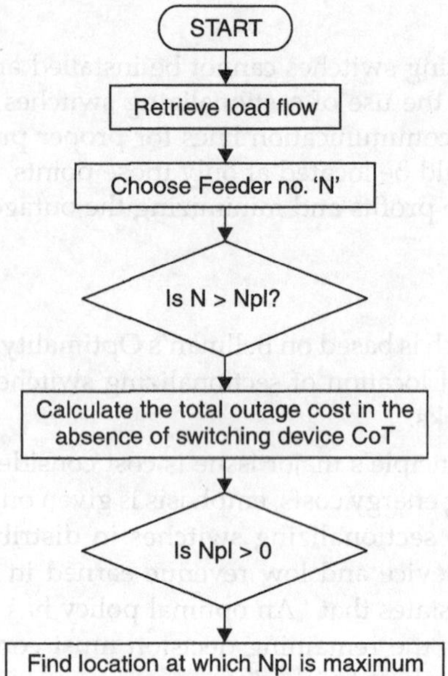

FIGURE 14.1 **Flow chart showing optimality theorem**

Laboratory Model

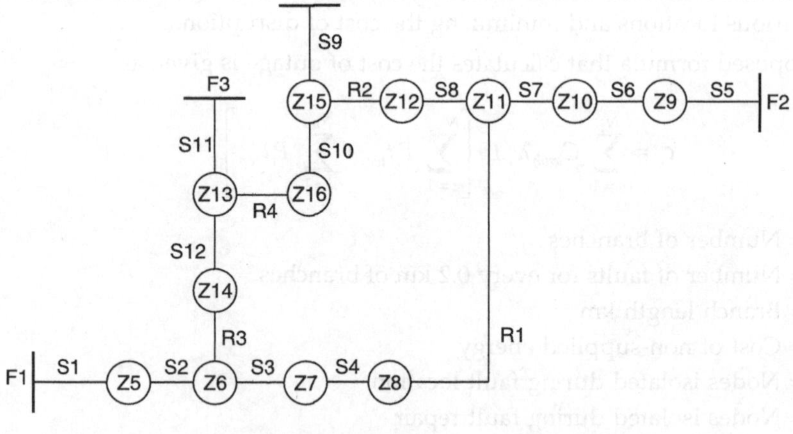

FIGURE 14.2 **Laboratory Model**

14.2.5 Results of the Laboratory Model

For a laboratory model, the following results are obtained. The number of switches in this case at a given zone will always be one, as the feeder model does not have any branches. However, in this case of a general model where there are '*n*' numbers of branches of the feeder, number and location of the placement of switches varies.

Result of Laboratory Model

Feeder	Zone	Length of Outage area	NetProfit/Loss
1	5	8.24 km	– 26047/-
1	6	6.99 km	123241/-
1	7	6.24 km	140791/-
1	8	4.08 km	144099/-
2	9	7.68 km	12699/-
2	10	6.53 km	91906/
2	11	5.68 km	109105/-
2	12	3.68km	113513/-
3	13	2.82 km	42936/-
3	14	1.77 km	86265/-
4	15	4.35 km	50317/-
4	16	3.35 km	74637/-

Each feeder has been analyzed individually at every zone or node location. The negative sign in the Net Profit/Loss column indicates a loss and hence it is not advisable to install a sectionalizing switch at that location. It can be seen that the maximum profit is obtained by placing the switch at node 8 for feeder 1 and then at node 7. Hence taking into consideration the money allocated for protection of distribution system, various sectionalizing switches can be installed.

Since the model does not have any branches it is a little easier to determine the outage cost. In a general case consisting of a number of branches, the outage cost would be increased; also, the node locations for positioning the sectionalizing switches would also increase. Both the programs can be developed.

14.3 CONCLUSION

The above proposed strategy reduces computational effort, as the location of second switch (in case of general program) depends on the location of the first switch *i.e.*, the initial stage is known. The proposed algorithm guarantees the location of devices on the main stem of distribution system, thereby obtaining maximum benefits; the location at remote ends is automatically eliminated. It is very useful for planning a distribution system especially due to the increased cost of energy, as it is very essential now-a-days to reduce the cost of outage.

LONG TYPE QUESTIONS

1. What are the various types of sectionaliser switches and their ratings?
2. Briefly elaborate on the benefits of using sectionaliser switches.
3. Explain Bellman's Optimality Principle.
4. Draw a flow chart explaining Bellman's principle.

MULTIPLE CHOICE QUESTIONS

1. High Voltage distribution systems use
   (*a*) Air Break Switches (*b*) Load Break Switches
   (*c*) MCCBs (*d*) Vacuum Circuit Breakers.
2. Any low voltage radial distribution network is vulnerable to
   (*a*) Faults (*b*) No load
   (*c*) Overload (*d*) None of these.

REFERENCE

1. *M. Tech. Project Reports* Guided by Prof. M.K. Khedkar.

Chapter 15 COMMUNICATION TECHNOLOGY

There are many communication methods available. Evaluation of different communication systems for data communication between Distribution control centre (DCC) and any point on the distribution network is required at the planning stage. The fundamental requirements for communication infrastructure are:

(*i*) Determination of system average message rate;

(*ii*) Capacity to handle the requisite amount of data and multitasking;

(*iii*) Data throughput and system response time should meet various application requirements;

(*iv*) It should allow for network growth and added applications.

The communication may be used individually or combined.

15.1 DIFFERENT TECHNIQUES USED FOR COMMUNICATION

Different techniques used for communication are Fibre optics, public telephone communication (PTC), Satellite Communication and Radio communication.

15.1.1 Fibre Optics

Fibre optics, with its explicit downward cost trend in terms of product as well as installation and maintenance costs, has become a widely accepted choice, as it offers both technical and commercial advantages over conventional systems that use metallic cables and radio links. The communication basically consists of a transmitter, and a receiver for information signals coming from the user's device, which is connected through copper wires to the switching centre or exchange, where it is changed into a digital signal like 1s or 0s for easy handling .The signal is then transferred to the transmitter. In the transmitter, the information signal which is electrical, drives an optical source: Laser or light emitting diode (LED), which in turn, optically modulates the information signal, which gets coupled to the optical fibre. The receiver located at the other side of the link detects the original signal and demodulates or converts back the optical signal to the original information signal (electrical). The signal is then connected through copper wires to the switching device or exchange for selection and connection to the proper user or user device. This type of communication is important for trunk routes power transmission lines. In such a situation, the problem of electrical interference is not encountered. This has advantage of high data rate (9600 bits and much more) and immunity from noise.

Optical fibres are available in cable form. Such cables can be laid in cable trenches. There is a trend to associate optical fibres with power cables or conductors themselves. Optical fibres are embedded within the cables/conductors, either earth or phase conductor. Optical fibre communication is expected to play an important role in future protection signalling, either as a dedicated protective relaying link, or as channels within a multiplexed system which carry all communication traffic such as voice, tele-control, tele-metering and protection signalling. Typical fibre optic communication system for underground distribution system for metropolitan area is shown in Fig. 15.1.

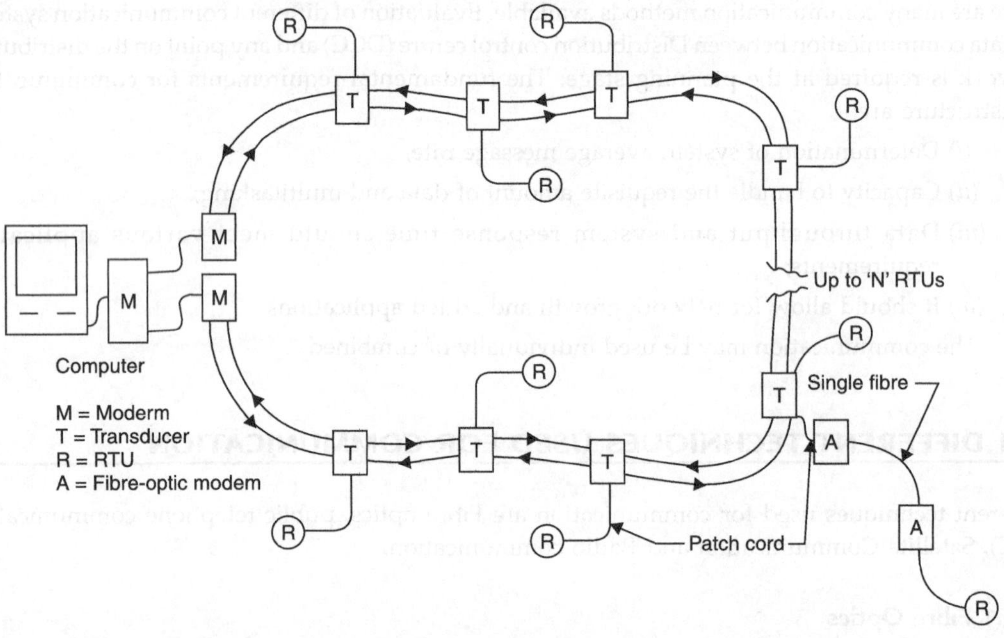

FIGURE 15.1 **Typical fibre-optic communication system for underground distribution system for metropolitan area**

15.1.2 Public Telephone Communication (PTC)

Dial-up and dedicated leased telephone lines are often used for Distribution Automation. Telecommunication services through Packet Switching Network (PSN), cellular radio or rural radio are viable, and may have the advantages of providing services in otherwise inaccessible places.

15.1.3 Satellite Communication

This technology is useful for load management controls. Satellite based VSAT (Very Small Aperture Terminal) network communication used for power systems is shown in Fig. 15.2. It has a 99.5% reliability factor A communication signal is sent from a VSAT to the satellite at a particular frequency. The signal is amplified, shifted to a different frequency and sent back to the VSAT at the destination. A VSAT network is distance insensitive. One has to pay only for the cost of VSATs and service charges, which are very reasonable. The whole network can be monitored and controlled centrally. If there is any problem in the network, it can be diagnosed from the hub. Primarily, there are three types of technologies for VSAT networks. The Government of India has approved the 'star' technology. In this configuration, at the centre of the network is a big earth station known as the hub with a dish diameter of 6-11 meters. The control of the network is at the hub and it maintains a communication link between VSATs scattered all over the country or region. The configuration is therefore called the Star.

The other technology is the 'MESH' technology. The configuration permits a group of VSATs to communicate directly with other VSATs going through the central hub. In this type of configuration, anyone of the VSATs functions as a hub and controls the whole network. This configuration enables better voice and video communication than the Star configuration. This technology has also been introduced in India. In the third type of configuration, there is point to point connection between the VSATs. In such a network; data, voice and video signals are carried directly between the locations.

In satellite communication, an extended C-band is used for communicating the data.

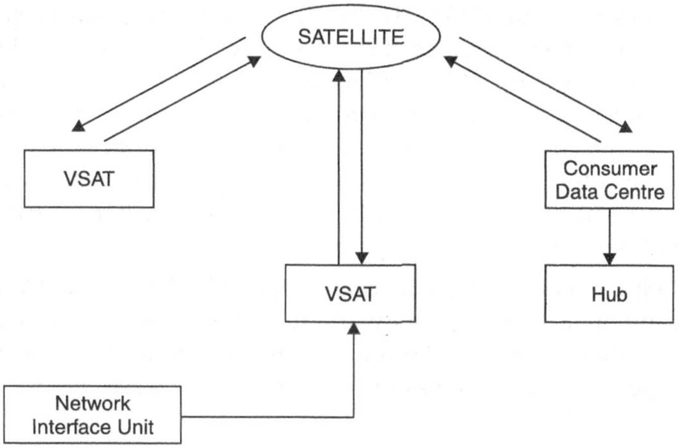

FIGURE 15.2 **Schematic diagram for VSAT Network**

15.1.4 Radio Communication

Radio communication systems, like PLC systems, are usually owned by the utility, but they have the important advantage of their operation being independent of the condition of the power system, furthermore, many data channels, including high speed channels, as well as voice communication can be provided. Radio technologies require licensing. Generally UHF and VHF radio is the most widely used medium. The disadvantages of radio are the limited channels available from licensing authorities, the relatively low data rate possible (typically 1200 bits/sec). The radio technologies generally used are:

(*a*) **UHF point to point Radio.** UHF radio in the 800-900 MHz range is widely used by utilities for two-way point to point communications. Systems in the 400 MHz range are also available. These systems usually support up to 15 voice and data channels operating at 4800 or 9600 bps. They require line of sight operation, in particular for the 900 MHz frequency bands. With flat terrain and antennas mounted at reasonable height, reliable operation over a range of 30 to 40 Km is practical. Dish antennas are used for point to point radios.

(*b*) **UHF Multi Address System Radio.** Multi address systems were developed as a substitute for dedicated telephone lines, and are widely used for SCADA/EMS/distribution automation application. They consist of master radio station which communicates with several remote radios with IRTU (RTU interfaced with distribution equipment). Each system uses a pair of frequencies (400 MHz –1000 MHz) one for master to remote communication and the other for remote to master, so that duplex communication is possible. Providing more remote sub master stations can increase coverage. An omni directional rod antenna is used at the master station. At the remote stations, directional antennas provide better performance, but less expensive, less conspicuous and vulnerable "stick" antennas are also used. This technology is useful for load management.

(*c*) **VHF Radio.** One way VHF radios (typically, the 154 MHz band widely used in the USA) is generally used to send load control commands. Inexpensive radio receivers/switches are available for giving load control switching commands.

(*d*) **Packet Switching Network (PSN).** A packet switching network uses low power UHF spread spectrum transmission (typical 900 MHz band) and does not require licensing of frequencies. It is a unique solution to frequency management using 'spread spectrum technology'. Due to its very low power density spectrum, it minimizes interference to existing services. The packet switching sends short burst of digital information over low power radios. These are data transmission networks with computerized nodes that perform specific communication tasks such as routing. Many power utilities in the world are using packet switching networks for many purposes, ranging from real time SCADA, EMS to line and capacitor switching, administrative data, and from inter utility to intra utility communications.

(*e*) **Cellular Radio.** This system is widely used for AMR, DSM and DA. Its benefits include wide area broadcast and monitoring capability, and it is ideal for field operations. The system has two levels of communication digital cellular for wide area network (WAN) at 900 MHz range and a low power spread spectrum RF as well as PLC channel for local area network (LAN).

15.2 DIFFICULTIES IN IMPLEMENTING COMMUNICATION TECHNOLOGIES

In this section, several practical examples are given to exemplify the communication difficulties faced by distribution automation.

Information Routing Difficulty—A Utility-Related DA Example

As shown in Fig. 15.3, the structure of a typical automated feeder (feeder A or B) consists of an ordinary feeder together with several Line Switches (LS) used to segment the feeder into sections. These LS are remotely controllable by a device called RTU (Remote Terminal Unit) located in the substation through another device called FTU (Facility Terminating Units) located besides the LS. Besides providing access to the control of the LS, these FTU's also report feeder conditions (currents and voltages) back to the RTU during normal periods. Today, most implementations of such system use a polling protocol between the RTU and the TUS while at all times *the R7T.J exercises full* control over the communication link. The purpose of sectionalizing the feeder with remote controllable LS's is to achieve a DA function, which is commonly known as FDIR (Fault Detection, Isolation, & Restoration). For example, suppose that a fault occurs on Section 2 of Feeder A in Fig. 15.3, CB 1 will trip. When this happens, CB1 also informs RTU-A that a fault has occurred. On the feeder an RTU-A starts to check each FTU if it has seen an over current. The fault section can be easily identified by RTU-A when it found that FTu-a2 saw an over current, while FTU-a3 did not. RTU-A then commands FTU-a2 and FTU-a3 to open their corresponding LS and after that, re closes CB 1.

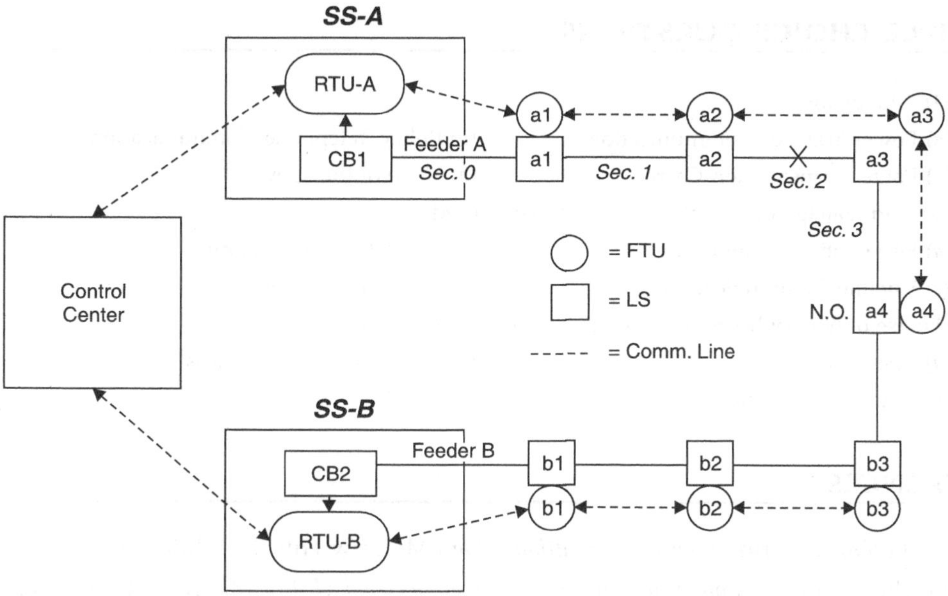

FIGURE 15.3 **Structure of typical automated feeder**

To further reduce the fault-effected region, another DA function viz. Load Transferring (LT) function is called. By checking network database and on-line load data, LT knows that Section 3 of feeder A can be transferred to Feeder B. It then commands RTU-A to close the normally open LS-a4. When this is completed, the fault-effected region will be minimized by this FDR-LT process. The load connected between a4 and a3 is energised theory Feeder B. A problem occurs after load transferring: when LS-a4 is closed, FTU-a4 and LS-a4 becomes part of feeder B. However, since the communication link to FTU-a4 remains unchanged, a method must be developed so that data can be routed between FTU-a4 and RTU-B. Under current system design, the routing must be done at the control centre (CC) with a special routing program, which can only route FDIR information. If other DA functions need to be implemented, the whole communication system would have to be totally re-studied or redesigned. Also, if different vendors are involved in the automation, CC may also have to do protocol or address conversion. All these make the integration of DA functions very difficult.

LONG TYPE QUESTIONS

1. What are the fundamental requirements of a communication system?
2. What are the different techniques used for communication?
3. Explain satellite communication.
4. Give the structure of a typical feeder.

MULTIPLE CHOICE QUESTIONS

1. PTC stands for
   (*a*) Power Transfer Communication (*b*) Public Telephone Communication
   (*c*) Power Transmission Carrier (*d*) None of the above.
2. For load management control, the technology used is
   (*a*) Fibre optic Communication (b) Radio Communication
   (*c*) Satellite Communication (*d*) None of the above.
3. The use of fibre optics is widely preferred nowadays because
   (*a*) Less cost (*b*) Signal attenuation is less
   (*c*) Has high data rate (*d*) All of the above.

REFERENCES

1. A.S.Pabla, "*Electrical Power Distribution*", Tata McGraw Hill, 4th Edition, 1998.
2. Newburry, John, "*Communication Field Trials for Total Utility Metering*" IEEE Transactions on Power Delivery Vol. II. No. 2 April 1996, P690.
3. Gelbebeln Lawerence, "*Distribution Automation Increase Reliability*", Transmission and Distribution World, October 1996, PP 38-41.
4. Pao-Hsiang His, Sh-lin Chen, "*Distributed Automation Communication Infrastructure*" IEE Transactions on Power Delivery, Vol. 13, no. 3, July 1998.
5. www.ieee.org

Appendix A INTRODUCTION TO ARTIFICIAL INTELLIGENCE BASED SYSTEMS

A.1 WHAT IS INTELLIGENCE?

Intelligence is the ability to learn, understand and think about things. It is a general mental capability that involves ability to reason, plan, solve problems, think abstractly, comprehend ideas and learn.

Scientifically, intelligence is the ability of a system to arrive at the most appropriate output inspite of all odds in a variety of situations. Typical situations are those where insufficient inputs, wrong inputs or noisy inputs may be present. There may be faults within the system itself or in the environment. Intelligence is the ability of the system to surmount these difficulties and produce appropriate output.

A.2 WHAT IS ARTIFICIAL INTELLIGENCE (AI)?

Artificial intelligence can be defined as intelligence executed by a manufactured or man- made system. It is an attempt to make a machine behave in ways that would be called intelligent if a human were to do the same thing.

AI is a diverse field where researchers pursue a wide range of problems, use a variety of methods and pursue a spectrum of scientific goals.

The final or ultimate aim of an artificial intelligence based system is to attain a human level of intelligence.

A.3 THE PARADIGM SHIFT

There was a paradigm shift in the conceptualizations of data and systems after the advent of the idea of Artificial Intelligence as is evident from the following table:

Table A.1 Comparison between everyday nomenclature and nomenclature in AI

Conventional Nomenclature	AI Nomenclature
Data	Pattern / Knowledge
Program / Algorithm	Reasoning / Planning / Problem Solving
GIGO(Garbage In Garbage Out)	Garbage In Useful Out
Deterministic	Deterministic / Probabilistic
Hard Wired	Evolution / Mutation / Adaptation
Serial	Parallel
Complex Architecture	Simple Architecture

A.4 THE AI TIME LINE

- **Pre-1940:** Von Hemholtz, Mach, Pavlov, etc. laid down general theories of learning, vision and conditioning. No specific mathematical models of neuron operation had yet been formulated.
- **1940s:** Hebb, McCulloch and Pitts studied the mechanism for learning in biological neurons and discovered that neural-like networks can compute any arithmetic function.
- **1950s:** Rosenblatt, Widrow and Hoff formulated the first practical networks and learning rules.
- **1960s:** Minsky and Papert demonstrated the limitations of existing neural networks. They claimed that new learning algorithms were not forthcoming and this resulted in some research being suspended.
- **1970s:** Amari, Anderson, Fukushima, Grossberg, Kohonen contributed to further progress, although at a slower pace.
- **1980s:** The research of Grossberg, Hopfield, Kohonen, Rumelhart, etc. led to important new developments and caused resurgence in the interest in the field.

A.5 THE AI TECHNIQUES

An artificial intelligence based system can be modeled on one of the following techniques:

1. Artificial Neural Networks
2. Fuzzy Logic
3. Genetic Algorithm

A system that incorporates more than one or all of the above mentioned techniques is called an Expert System.

A.6 ARTIFICIAL NEURAL NETWORKS

A neural network is a hugely distributed parallel processor made up of a number of processing units, which have a natural tendency to store knowledge and making it available as and when required. The simplest such processing unit is called a neuron.

The inspiration for ANN is drawn from the human brain; a highly complicated structure made up of neurons and can perform or be trained to perform certain computations many times faster than most computers in existence today. Some examples of such computations include pattern recognition, motor control, etc.

Neural networks are thus, machines that can be modeled to perform a particular task and in doing so, mimicking the human brain.The network can be modeled by electronic components or simulated on computer software.

Thus, the neural network resembles the human brain in the following aspects:

Knowledge is acquired by the network from its environment through a learning process.

Inter-neuron connection strengths, called synaptic weights are used to store the acquired knowledge. The procedure used for training the network is called a learning algorithm, whose function is to modify the synaptic weights of the network to attain the desired objective.

A.6.1 The Human Brain as a Neural Network

The Human Nervous System may be viewed as a three stage system.

The central part of the system is the brain which is represented by the neural net which continually receives information, processes it and makes appropriate decisions. Forward and backward transmission of information is possible between these stages. The receptors convert stimuli from the human body or the external environment into electrical impulses. Receptors can be viewed as electrical transducers which pass on information to the next stage, the neural net.

The neural net after processing the information passes it on to the effectors which convert the electrical impulses into discernible system outputs.

Ramon Y Cajal introduced the idea of neurons as structural constituents of the brain in the year 1911. After this pioneering work, the study of the brain has been much easier. Neurons are 5–6 times slower than silicon logic gates; events in a silicon chip happen in the nanosecond range while neural events happen in the millisecond range. What makes the brain so fast then? It is because of the large number of neurons (approx. 10 billion) that are present in the brain. Thus the human brain is a very efficient structure. Specifically, the energy efficiency of the brain is approx 10^-16 J per operation per second. The corresponding value for the best computers in use today is 10^-6 J per operation per second. A typical neuron is shown in Fig. A.1.

It consists of the axon, the dendrites and the synapses. Axons are the transmission lines which transmit information from one neuron to another. Dendrites are the receivers which constitute two types of self elements that are morphologically distinguished. Synapses are elementary functional units that mediate the interaction between the neurons. The body of the neuron is called the 'soma'. Thus the information flows from the dendrite to the soma. It is processed in the soma and then transmitted through the axon.

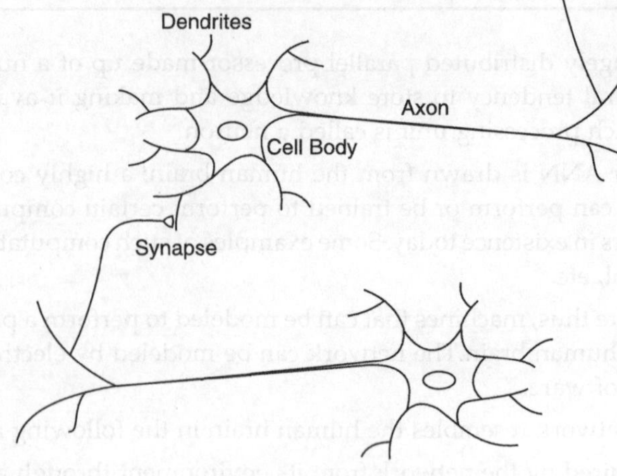

FIGURE A.1 **A Typical Neuron**

A.6.2 Neural Models and Architectures

The commonly used models are: the single input model and the multiple input model as shown in Figs. A.2 and A.3.

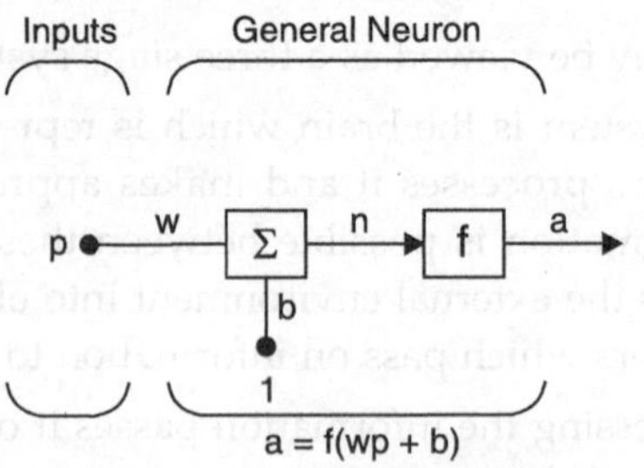

FIGURE A.2 **Single Neuron Model**

As shown in Fig. A.2, the single neuron model consists of a single input that gets multiplied by a weight (*w*). This is processed by the summation block and the transfer function to give the output '*a*' = *f*(*wp* + *b*) . '*b*' is called the bias, the input to which is always unity.

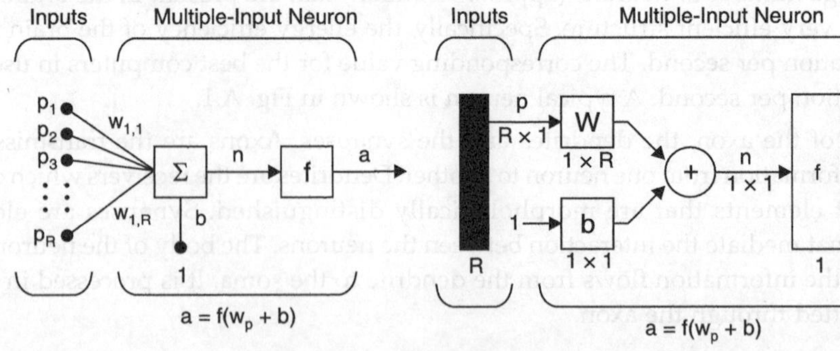

FIGURE A.3 **The Multiple Input Neuron Model**

The multiple input neuron consists of a number of inputs p_i as shown in Fig. A.3. Each input is multiplied by a weight. The summation block adds the products of the weights and the inputs along with a bias whose input is always unity. This produces the output '*n*' which when passed through the transfer function block gives the output '*a*'. Thus, $a = f(Wp + b)$

Here, *p* refers to the input vector of dimensions $R \times 1$ given as:

$$\mathbf{p} = \begin{bmatrix} p_1 \\ p_2 \\ \vdots \\ p_R \end{bmatrix}$$

The weight matrix is a matrix of dimensions $S \times R$.

From the above models, three basic elements of the neuron model can be identified.

(I) A set of synapses each of which is characterized by a weight or strength of its own. Specifically, an input pi at the input of the synapse I connected to neuron j gets multiplied by the synaptic weight w_{ji}.

(II) An adder for summing the input signals weighted by the respective synapses.

(III) An activation function or transfer function for limiting the amplitude of the output of the neuron.

Common activation functions used are shown in Fig. A.4:

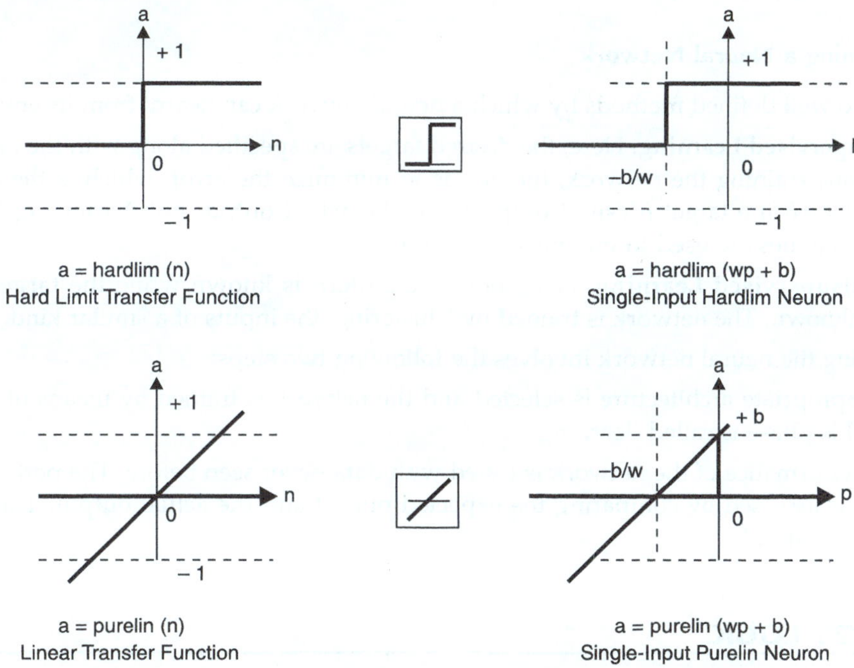

FIGURE A.4 **Commonly used activation (transfer) functions in a Neural Network**

A.6.3 Properties of Neural Networks

1. **Non-Linearity.** An artificial neuron can be linear or non linear. A neural network (interconnection of neurons) by itself is non-linear.
2. **Input Output Mapping.** Supervised learning involves modification of synaptic weights by applying a set of labelled training samples. An example is picked at random and the weights are modified to minimize the error. This process is repeated until there are no further changes in the weights. Thus the network learns from the examples by creating an input output mapping for the problem.
3. **Adaptivity.** A neural network trained to operate in a particular environment can be easily retrained to operate in a changed environment. In a non-stationary environment the network can be designed to change its weights in real time.
4. **VLSI Implementability.** The massively parallel nature of the neural network makes it suited for fast computation of certain problems.
5. **Fault Tolerance.** Due to the distributed nature of the information stored in the network, the damage to the network has to be extensive before the response of the network is seriously degraded. Thus a neural network exhibits a gradual and graceful decline in performance rather than a sudden failure.
6. **Neurobiological Analogy.** The design of a neural network is motivated by its analogy with the human brain, which is a living proof that fault tolerant parallel processing is not only physically possible, but also fast and powerful.

A.6.4 Training a Neural Network

There are two well defined methods by which a neural network can 'learn' from its environment:

1. **Supervised Learning.** Here, the desired targets are specified along with the pattern and while training the network, the aim is to minimize the error, which is the difference between the target (desired output) and the actual output got. Various optimization techniques are used to minimize this error.
2. **Unsupervised Learning.** Here, only the pattern is known while the target remains unknown. The network is trained by 'clustering' the inputs of a similar kind.

Training the neural network involves the following two steps:

An appropriate architecture is selected and the network is trained by means of a suitable algorithm. This phase is called 'learning'.

The performance of the network is tested with data never seen before. The performance of the network is assessed by comparing the expected output and the actual output. This phase is called 'generalization'.

A.7 FUZZY LOGIC

Fuzzy logic is an AI technique which models and incorporates the process control expertise of a designer or an operator. It accommodates reasoning that is approximate rather than precise.

The premise for fuzzy logic stems from several observations such as:

1. The knowledge possessed by human beings of any situation is generally imperfect.
2. Humans solve complex problems with approximate data.

Thus, rather than modeling the system, it is more useful to model the behaviour of the human operator controlling the system.

The formulation of fuzzy rules for a system has the following advantages:

1. Incorporation of numerous variables, whose effect would be difficult to model conventionally.
2. Incorporation of fuzzy feed forward techniques.

However, fuzzy logic cannot provide a solution when no one knows how the system operates, or people are unable to manually control it.

Fuzzy rules do not replace the conventional methods. Rather, they complement them.

The solving of a problem by the use of fuzzy logic involves the following steps:

- The fuzzy set is defined and membership values are defined for each element of the set.
- The 'crisp' input is fed to a 'fuzzifier' which converts this 'crisp' value into a fuzzy value.
- The output of the 'fuzzifier' is decoded as per the rules defined in i) above and is fed to a 'de-fuzzifier' which gets a 'crisp' value as output.

In a nutshell, fuzzy logic:

- Formalizes and mimics the expertise of a human operator in process control and tuning.
- Provides a simple method for controlling processes that are difficult to model.
- Has the ability to continually take into account exceptions of different kind and incorporate them into their expertise.
- Take into account the effect of several variables and perform a weighted merging of their effects.

A.8 GENETIC ALGORITHM

It is a procedure that helps us get an optimum result out of a particular situation. The solutions are watched using a fitness function. Thus the algorithm is stopped when it is found that the solution has attained a high degree of fitness.

The initial population used for evolving a solution is created randomly and so is the position of crossover. However, the process is not an entirely random one from the sense that there is a fixed objective, *i.e.*, obtaining a fitter solution.

The genetic algorithm and the steps involved as applied to a biological system and the corresponding mathematical operations are enumerated in Table A.2:

Table A.2 Comparison between biological and mathematical operations

Biological Inspiration	Mathematical Operation
Population	Collection of candidate/ tentative solution
Coding	Solution is not dealt with directly but in a coded form. The most popular form of coding is binary coding.
Selection	A fitter solution is selected using pre-defined criterion. This is called a fitness function.
Crossover-it is the swapping of the values of two genes at a random position.	Mimicked as it is
Mutation-it is a highly random process by which the 1's in the gene are altered to 0 and vice versa	Mimicked as it is

The various optimisation methods are:

- Calculus Based methods
- Direct methods
- Indirect methods
- Guided Random methods
- Simulated Annealing method
- Genetic Algorithm method
- Parallel method
- Serial method
- Simple Genetic Algorithm method
- Genitor method
- Genesis
- Enumeration
- Dynamic Programming

REFERENCES

1. Hagan: *"Fundamentals of Artificial Neural Networks"*.
2. Sources from the internet.

Appendix B EVOLUTIONARY COMPUTING

B.1 INTRODUCTION

Evolutionary computation used computational models of evolutionary processes as key elements in the design and implementation of computer based problem solving systems.

There are a variety of evolutionary computational models that have been proposed and studied, which are referred to as Evolutionary Algorithms (EA). They share a common conceptual base of simulating the evolution of individual structures *via* processes of selection and reproduction.

These processes depend on the perceived performance (fitness) of the individual structures as defined by the environment.

More precisely, evolutionary algorithms maintain a population of structures that evolve according to rules of selection and other operators such as recombination and mutation. Each individual in the population receives a measure of its fitness in the environment. Selection focuses attention on high fitness individuals, thus exploiting the available fitness information. Recombination and mutation perturb those individuals, providing general heuristics for exploration.

Although simplistic from a biologist's viewpoint, these algorithms are sufficiently complex to provide robust and powerful adaptive search mechanisms.

An evolutionary algorithm typically initializes its population randomly, although domain specific knowledge can also be used to bias the search. Evaluation measures the fitness of each individual according to its worth in some environment. Evaluation may be as simple as computing a fitness function or as complex as running an elaborate simulation.

Selection is often performed in two steps namely parent selection and survival. Parent selection decides who become parents and how many children the parents have. Children are created by recombination, which exchanges information between parents, and mutation, which further perturbs the children. The children are then evaluated. Finally, the survival step decides who survives in the population.

B.2 BIOLOGICAL BASIS

Evolution in nature or anywhere else is not a purposive or directed process, *i.e.*, there is no evidence to support assertion that the goal of evolution is to produce mankind. Indeed, the processes of nature boil down to a haphazard *generation* of biologically diverse organisms. Some of the evolutions are determined by natural *selection* of different *individuals* competing for resources in the *environment*. Some are better than others. Those that are better are more likely to survive and propagate their genetic material. In nature, the encoding for genetic information (*genome*) is done in a way that admits asexual *reproduction*. Asexual reproduction typically results in offspring that are genetically identical to the *parent*.

Sexual reproduction allows some shuffling of chromosomes producing offspring that contain a combination of information from each parent. At the molecular level, a pair of almost identical chromosomes bump into one other, exchange chunks of genetic information and drift apart. This is the recombination operation, which is often referred to as crossover because of the way biologists have observed strands of chromosomes crossing over during the exchange.

Recombination occurs in an environment where the selection of who gets to mate is largely a function of the fitness of the individual. Some *luck* is usually involved too.

Some evolutionary algorithms use a simple function of the fitness measure to select individuals probabilistically to undergo genetic operations such as crossover. This is fitness proportionate selection. Other implementations use a model in which certain randomly selected individuals in a sub-group compete and the fittest is selected. This is called tournament selection.

Much EA research has assumed that the two processes that contribute most to evolution are crossover and fitness based selection. Evolution by definition absolutely requires diversity in order to work. In nature, an important source of diversity is mutation. In an EA, a large amount of diversity is usually introduced at the start of the algorithm by randomizing the *genes* in the *population*. The importance of mutation which introduces further diversity while the algorithm is running therefore continues to be a matter of debate.

It cannot be stressed too strongly that an evolutionary algorithm as a simulation of a genetic process is not a random search for a solution to a problem (highly fit individual). EAs use stochastic processes but the result is distinctly non-random, which is better than random.

B.3 VARIETIES OF EVOLUTIONARY ALGORITHMS

The methodologies of Evolutionary Algorithms that have emerged in the last few decades are :

1. Evolutionary Programming
2. Genetic Algorithms

Although similar at the highest level, these methods implement the EAs differently.The differences touch upon almost all aspects of EAs, including the choices of representation of the individual structures, types of selection mechanisms used, forms of genetic operators, and measures of performance.

These approaches have in turn inspired the development of additional EAs like Classifier Systems, the LS systems, Adaptive Operator Systems, messy Gas, etc.

B.4 GENETIC ALGORITHMS

Genetic Algorithms (GAs) developed by Holland in 1975 have traditionally used a more domain independent representation, namely, bit strings. However many recent applications of GAs have focused on other representations, such as graphs, Lisp expressions, ordered lists and real valued vectors.

The remaining details about Genetic Algorithms have been discussed in Appendix A and Chapter 12.

B.5 EVOLUTIONARY PROGRAMMING

Evolutionary Programming (EP), originally conceived by Lawrence J. Fogel in 1960, is a stochastic *optimization* strategy similar to Genetic Algorithms , but instead places emphasis on the behavioural linkage between *parents* and their *offspring* , rather than seeking to emulate specific *Genetic Operators* as observed in nature. Evolutionary Programming is similar to *Evolutionary Strategies*, although the two approaches developed independently.

Like both ES and GAs, EP is a useful method of optimization when other techniques such as gradient descent or direct , analytical discovery are not possible. Combinatorial and real-valued *function optimization* in which the optimization surface or *fitness* landscape is *rugged*, possessing many locally optimal solutions, are well suited for evolutionary programming.

The Process:

For EP, like GAs , there is an underlying assumption that a fitness landscape can be characterized in terms of variables , and there is an optimum solution(or multiple such *optima*) in terms of those variables.The basic EP method involves 3 steps (Repeat until a threshold for iteration is exceeded or an adequate solution is obtained):

1. Choose an initial *population* of trial solutions at random. The number of solutions in a population is highly relevant to the speed of optimization, but no definite answers are available as to how many solutions are appropriate (other than 1) and how many solutions are just wasteful.
2. Each solution is replicated into a new population. Each of these offspring solutions are mutated according to a distribution of *mutation* types, ranging from minor to extreme with a continuum of mutation types. The severity of mutation is judged on the basis of the functional change imposed on the parents.
3. Each offspring solution is assessed by computing its fitness. Typically, a stochastic tournament is held to determine N solutions to be retained for the population of solutions, although this is occasionally performed deterministically. There is no requirement that the population size be held constant, however, nor that only a single offspring be generated from each parent.

It should be pointed that Evolutionary Programming typically does not use any *crossover* as a genetic operator.

B.6 EVOLUTIONARY PROGRAMMING AND GENETIC ALGORITHMS

There are two important ways in which EP differs from GAs. The typical GA approach involves encoding the problem solutions as a string of representative tokens, the *genome*. In EP, the representation follows from the problem. A neural network can be represented in the same manner as it is implemented, because the mutation operation does not demand a linear coding.

Second, the mutation operation simply changes aspects of the solution according to a statistical distribution which weighs minor variations in the behaviour of the offspring as highly probable and substantial variations as increasingly unlikely. Further, the severity of mutations is often reduced as the global optimum is approached. Several techniques have been proposed and implemented which address this difficulty, the most widely studied being the *MetaEvolutionary* technique in which the variance of the mutation distribution is subject to mutation by a fixed variance mutation operator and evolves along with the solution.

REFERENCE

1. Sources from the internet.

Appendix C

BASIC EQUIPMENTS IN DISTRIBUTION SYSTEM

C.1 SWITCHGEAR AND CIRCUIT BREAKER

C.1.1 Introduction

Circuit breakers are one of the most important devices used in a power system whose primary function is to protect the system, thus ensuring continuity of supply.

Of all the equipments, employed in a power system for generation, transmission and distribution of electrical energy, circuit breakers are the least utilized devices. However, they are responsible for continuously monitoring the status of the system and operating in case of an anomalous situation.

Every electrical system is susceptible to faults. These faults may be caused by incorrect switching, short circuits between two or more lines or between line and ground, deterioration of insulation, etc. Such conditions result in the flow of currents of large magnitude, as a result magnifying heating and electromagnetic forces, both of which are proportional to the square of current.

In inter-connected systems, the transient stability of the system is disturbed if the fault currents are allowed to remain on the system for long periods.

From the above discussion, the basic requirements of a circuit breaker can be deduced as:

1. The breaker must be capable of continuously carrying rated system current at the point of installation.
2. It must be capable of breaking the circuit under abnormal conditions.
3. The operation of the circuit breaker must not affect or damage other equipment present in the system.

The term 'switchgear' is used for the entire assembly consisting of the circuit breaker and the control apparatus.

C.1.2 Location

Switchgears are located in places where isolation and protection are required. These locations include generators, transformers and sub-stations. In a sub-station, switchgear is located on both, the high voltage as well as low voltage sides.

C.1.3 Principle of Operation

When the contacts of a circuit breaker separate, an arc is struck between the contacts. The circuit breaker thus has to be designed in such a way that this arc is controlled and eventually extinguished so that no damage is caused to equipment or personnel.

Two methods are available for extinguishing the arc:

1. *High Resistance Method*: Here, a constantly increasing resistance is inserted in the circuit and simultaneously, the arc is cooled, so as to reduce current to a value that is insufficient to maintain the arc. This is done by increasing the length and resistance of the arc. This principle is employed in d.c. circuit breakers.
2. *Zero Current Method*: This method is applicable only to a.c. circuit breakers. The arc is broken at the instant when the current passes through zero (this zero crossing of current occurs twice in one period).

C.1.4 Types of Switchgear

Based on the housing, switchgear can be classified as:

1. *Indoor type*: Used for voltages up to 33 kV and when large currents are not involved.
2. *Outdoor type*: They are used for higher voltage and current ratings. Since they are placed outdoors, they have to be designed to withstand rain, wind etc. As a result, their physical dimensions are larger than that of an indoor circuit breaker.

Based on the type of circuit breaker used, switchgear can be classified as:

- *Oil Circuit Breaker*: Here, the contacts are surrounded by insulating oil and they are used to open and close high voltage circuits carrying relatively large currents. When the contacts are drawn apart, the oil covering them tends to quench the arc by its cooling effect and by the gases thereby generated, which tend to blow out the arc. Since oil has better insulating properties than air, distance between live parts as well as live parts and ground can be reduced without the possibility of arcing.

  Thus the oil serves a two-fold purpose:

  (*a*) Extinguishing the arc.

  (*b*) Providing insulation between the live parts and the metallic tank.

  The properties desired of oil used in a circuit breaker apparatus are enumerated below:

  (*i*) High dielectric strength.

  (*ii*) Free from acid, alkali or sulphur.

  (*iii*) Low viscosity.

(*iv*) Low freezing point.

(*v*) Emulsion resistant.

- *Air Blast Circuit Breaker*: The most common method of arc control in air blast circuit breakers is that of subjecting the air to a high pressure blast. A puff of air or magnetic field is used to blow the arc into arc-chutes, where the arc gets elongated and is no longer able to sustain itself. Air blast circuit breakers are of two types, axial and cross blast, depending on the direction of the air blast with respect to the arc.

  The advantages offered by air blast circuit breakers over other types are:

  (*i*) Elimination of fire hazard.

  (*ii*) Uniform high speed operation independent of magnitude and power factor of interrupted current.

  (*iii*) Easy maintenance.

- *Gas Circuit Breakers*: This type of circuit breaker uses an inert, nontoxic, odourless gas *i.e.*, sulphur hexafluoride for insulation. This gas can extinguish arcs rapidly. It has excellent heat dissipating characteristics, and its dielectric strength is much greater than that of oil. Installation and maintenance requirements of these breakers are less costly than for other types.

  There are two different techniques using SF_6

  (*i*) Two pressure circuit breakers, in which the gas is permanently stored at the pressure required for interruption, arc extinction being achieved by the expansion of the high pressure gas towards the low pressure enclosure.

  (*ii*) Single pressure type, in which the pressure required for arc extinction is got by the movement of the contacts, by displacement of a piston inside a cylinder.

  The two pressure technique promises higher performance levels, but is less economical compared to the single pressure type.

- *Vacuum Circuit Breaker*: When the contacts separate, the current to be interrupted initiates a metal vapour arc discharge and flows till the next current zero. The arc is then extinguished and the metal vapour condenses on the metal surface in a very small time. The rapid build up of dielectric strength in the space between the contacts enables the arc to be extinguished even if contact separation occurs immediately prior to current zero. Also the arc voltage developed in VCB is low due to the high conductivity of the metal plasma. Use of VCB is especially suited in industrial applications, where switching frequency is high combined with a high degree of pollution.

C.1.5 Operating Mechanism

The trip mechanism of the breaker must be able to operate in a matter of milliseconds. Considerable mechanical stresses are involved, and thus the design of the breaker, apart from being simple, must be robust and reliable.

The main factors to be taken into account while choosing an operating mechanism are:

(*i*) Small power consumption.

(*ii*) Quick movement to overcome the repulsive forces associated with the closing of the breaker contacts.

(*iii*) High initial acceleration, which reduces smoothly after the contacts touch so as to avoid the use of heavy dampers.

On basis of the above points, the operating mechanisms may be classified as:

(*i*) Hand operation.

(*ii*) Motor operation.

(*iii*) Solenoid operation.

(*iv*) Motor operated compressed spring operation.

(*v*) Pneumatic operation.

C.2 FUSES

C.2.1 Introduction

An electrical fuse is a current interrupting device which protects an electrical circuit in which it is installed by creating an open circuit condition in response to excessive current. The current is interrupted when the element or elements which carry the current are melted by heat generated by the current. Fuse terminals form an electrical connection between an electrical power source and an electrical component or a combination of components arranged in an electrical circuit. A fusible link is connected between the fuse terminals, so that when electrical current flowing through the fuse exceeds a predetermined limit, the fusible link melts and opens the circuit through the fuse to prevent electrical component damage.

Fuses are used on power systems up to 115,000 volts AC. High-voltage fuses are used to protect instrument transformers used for electricity metering, or for small power transformers where the expense of a circuit breaker is not warranted. The fuses are arranged to disconnect the faulted equipment or circuit promptly from its source of supply before damage can occur. Fuses are used extensively in high voltage electrical networks in order to protect the electrical equipment in the network from damage caused by surges through the system, generally due to short-circuits or overloads.

Comparison between Fuse and Circuit-breaker

- Fuses have the advantages of being less costly and simpler than a circuit breaker for similar ratings. For example, in distribution systems, a power fuse may be used to protect a transformer serving 1–3 houses. A circuit breaker at 115 kV may cost up to five times as much as a set of power fuses, thus increasing the capital costs.
- Current-limiting fuses operate so quickly that they limit the total "let-through" energy that passes into the circuit, helping to protect downstream equipment from damage.

These fuses clear the fault in less than one cycle of the AC power frequency. Circuit breakers cannot offer similar rapid protection.

- Circuit breakers which have interrupted a severe fault should be removed from service and inspected and replaced if damaged. Fuses, on the other hand have to be replaced once they operate.
- Circuit breakers must be maintained on a regular basis to ensure their mechanical operation during an interruption. This is not the case with fuses, in which no mechanical operation is required for the fuse to operate under fault conditions.

C.2.2 Parameters of Fuses

Following are some parameters of fuses:

1. *Rated current I_N*: This is the maximum current that the fuse can continuously pass without interruption to the circuit, or harmful effects on its surroundings.
2. *The I^2t value*: This is a measure of the energy required to blow the fuse element and is an important characteristic of the fuse. It is an indication of the "let-through" energy passed by the fuse which downstream circuit elements must withstand before the fuse opens the circuit.
3. *Breaking capacity*: The breaking capacity is the maximum current that can safely be interrupted by the fuse. Some fuses are designated High Rupture Capacity (HRC) and are usually filled with sand or a similar material.
4. *Voltage rating*: The voltage rating of a fuse should always be greater than or equal to the circuit voltage. Low-voltage fuses can generally be used at any voltage up to their rating. Some medium-voltage and high-voltage fuses used in electric power distribution will not function properly at lower voltages.

C.2.3 Types of Fuses

Following are some types of fuses:

- *Open or Link fuses*

  It is merely a piece of metal ribbon or fuse wire connected between two terminals. Since these fuses are open type, they tend to smudge or burn the surrounding surfaces, when they vapourise. It is easy to replace such type of fuse.

- *Enclosed fuses*

  The fuse element is enclosed within a tube of insulating material, such as glass, fiber, or mica. The tube is filled with a material, usually in powdered form, which absorbs the heat and liberated gases when a fuse is blown. They are usually used with potential transformers. Types of such fuses are:

  1. Liquid-Filled fuses
  2. Sand and Boric acid fuses
  3. Oil fuses

- *Expulsion fuses*

  In this type, the fuse wire is so designed that the expulsion of gases formed by the vapourising of the fuse and the expansion of the surrounding air is sufficient to blow out the arc.

C.3 CURRENT AND VOLTAGE TRANSFORMERS

CTs and PTs are the 'eyes and ears' of the protective system. They extract information from the power system and form an important link between the high voltage, high current power system and the low voltage, low current protective system. The two main functions of the CT and PT are:

(*i*) To electrically isolate the protective relays from the high voltage power system to ensure safety of operating personnel.

(*ii*) To step down the current and voltage to standard values so that design of relay can be standardized irrespective of the actual primary voltage and current.

Voltage Transformers

Voltage transformers are necessary on every system for:

(*i*) Indication of voltage conditions.

(*ii*) Metering of the supply of energy.

(*iii*) Relaying.

(*iv*) Synchronizing.

Types of Voltage Transformers

(*i*) *Magnetic Type*: They work on the same principle as power transformers. However, the design is different for the two cases in the sense that the load to be borne by the PT is limited to a few hundred VAS at the most. The main objective in the design of a PT is to minimize the ratio and phase angle errors so as to ensure accurate measurement. Another significant distinction between a PT and a power transformer is that in case of a PT, only one terminal of the primary winding is brought out for connection to the line. The second terminal is earthed along with the core and steel housing.

(*ii*) *Capacitive Voltage Transformer*: It is used for system voltages greater than 132 kV. A major benefit offered by a CVT is that it enables measurement of line to earth voltage to be made with simultaneous provision for carrier frequency coupling, which has wide applications in modern high voltage networks for telemetering, remote control and telephone communication purposes.

CVTs are of two kinds—coupling capacitor type and bushing type.

Specification of a Voltage Transformer

(*a*) Ambient Temperature—40°C

(*b*) Altitude—up to 1000 m above sea level.

(*c*) Atmospheric Condition—not heavily polluted and not conductive to the growth of fungi and condensation of moisture.

(*d*) Rated Secondary Voltage—110 or 63.51 V. It is essential that the PT primary winding neutral be earthed so as to provide for the flow of zero sequence exciting current. If the primary winding is not earthed, an open delta winding may develop a voltage entirely by third harmonics. This voltage is not related to any residual voltage in the primary system.

Current Transformers

A CT can be considered as a step up transformer with independent primary and secondary windings. At normal operating conditions, rated line current flows through the CT primary. The secondary of the CT, connected to measuring instruments and relays, supplies a current proportional to and in phase with the primary current.

Theory

The CT is similar in construction to the single-phase power transformer and also obeys the same fundamental laws. The difference between the two lies in the fact that while in a power transformer, the primary current is controlled by power demand in the secondary circuit, in a CT, the primary current is imposed on the transformer by the supply system.

Secondary Current: 1 A or 5 A. For transformers intended for delta connection, these ratings divided by (root) 3 shall be considered as standard values.

Difference between Instrument and Power Transformers

The voltage in a transformer is given by $E = 4.44\phi_m fN$, where f is the frequency, N the number of turns and ϕ_m the maximum flux.

For a power transformer, an increase in load would lead to an increase in secondary current and as a result, primary current. This would call for the establishment of a large magnitude of flux. The operating point of a power transformer being at the knee point of the B-H curve, it will go into saturation.

In such a situation, the magnetizing current becomes very large and a stage is reached when all the transformed secondary current is used up in exciting the transformer core. Thus the transformer cannot deliver any current to the load. However, large voltage spikes are generated when the flux crosses zero.

A CT however, must faithfully reproduce not only the normal operating current, but in case of a fault must also faithfully transform the fault current so as to enable the relays and other protective devices connected to its secondary to detect the fault and perform the required corrective actions. Thus a CT must be designed to operate well below the knee point.

Safety Precautions

Care must be taken to ensure that the secondary of the CT is not disconnected from its load when current is flowing in its primary. In such a situation, the transformer secondary will attempt to continue driving current across the effectively infinite impedance. This will lead to the production of a very high voltage across the open secondary and may lead to arcing, thus compromising equipments and operator safety. This will also permanently affect the accuracy of the transformer.

C.4 LIGHTNING ARRESTORS

Basic Insulation Level

We need to ensure that the flashover strength of all insulation in the sub-station exceeds a certain pre-defined level. Thus it becomes necessary for us to define a common insulation level for all equipment operating at a particular system voltage.

The basic insulation level of the equipment determines the principal dielectric qualities of the apparatus and is expressed for impulse tested apparatus for 1.2/ 50 ms impulse wave.

Types of Protective Devices

- *Rod Gap*: This is the simplest protective device available. However, it is rarely used, as it does not cut off power voltage after it has been flashed over by a surge. This would imply a short circuit on the system every time a flashover occurs across the rod gap as a result of a surge. Moreover, the breakdown characteristics of the rod gap are affected by atmospheric conditions and the polarity of the incident wave. In view of this, spark gaps are generally used as a back up protection in case where the lightning arrestor gets disconnected from the line.
- *Expulsion Type Lightning Arrestor*: This device consists of a spark gap enclosed in a fibre tube and another external spark gap in air, in series. On occurrence of a high voltage, the two spark gaps breakdown, thus establishing a conducting path in the form of an arc. When the arc touches the walls of the fibre tube in passing down, it vapourises a small part of the material. The gas thus produced drives out the air ionized by the arc and when the current passes through zero, the arc is extinguished.

  These arrestors have better volt-time characteristics than rod gaps. They are extensively used to prevent flashover of transmission line insulators.
- *Valve Type Lightning Arrestor*: This device consists of a divided spark gap in series with a resistance element having a non-linear characteristic. On account of breakdown due to the presence of high voltage, the device becomes conductive. The non-linear resistor, which offers low resistance to the flow of high currents and high resistance to the flow of power frequency follow currents, aids in breaking the follow current.

  The resistance elements are made of silicon carbide arranged in the form of small cylindrical blocks bound together by an inorganic binder.

Lightning Arrestor Specifications

1. **Rated Voltage:** It is the designated maximum permissible r.m.s. value of power frequency voltage that an arrestor can support across its line and earth terminals without getting

damaged. The rating of an arrestor must be higher than the normal line to ground voltage, otherwise it may draw a large follow current leading to thermal overloading and failure.

The higher system voltage (HSV) is deemed to be 110% of the corresponding nominal system voltage.

2. **Effectively earthed System:** A system is said to be effectively earthed if under any fault condition, the line to earth voltage on the healthy phases will not exceed 80% of the system line-line voltage.
3. **Non-effectively Earthed System:** A system is said to be non-effectively earthed if under earth fault condition, the line-earth voltage on the healthy phases is more than 80% but does not exceed 100% of the line-line voltage.
4. **Isolated System:** In such a system, the neutral is not grounded and line-line voltage of a healthy phase may exceed 100% of line-line voltage in the event of a ground fault on one phase.
5. **Nominal Discharge Current:** It is the discharge current having a designated crest value and waveshape. The waveshape specifies for these ratings is 8/20 μs.
6. **Rated Frequency:** The rated frequency is 50 Hz.
7. **Maximum Impulse Sparkover Voltage:** It is the amplitude of 1.2/50 μs voltage wave on which the arrestor sparks over 5 times out of 5.
8. **Residual Voltage:** It is the maximum value of the voltage appearing between the terminals of the arrestor at the time of the discharge of a surge current wave. This value depends on the type and nominal voltage of the arrestor, the wavefront gradient and the amplitude of the discharge current.
9. **Maximum Discharge Current:** It is the maximum value of discharge current arrestor can pass without damage or modification of its characteristics. The waveshape referred to here is 4/20 μs.
10. **Cut off Voltage:** It is the highest power frequency r.m.s. voltage the arrestor can withstand across its terminals while still being capable of interrupting the follow current effectively.

C.5 CO-ORDINATION AMONGST SUB-STATION EQUIPMENT

The major sub-station equipment namely transformers, breakers, isolating switches and instrument transformers are generally manufactured for the same insulation level.

Four levels of insulation in a station are recognized: the bus insulation is highest, followed by breakers, transformers and lightning arrestors in that order. The percentages of margins between these levels is not uniform but can be attained on a reasonably constant basis by the use of standard equipment commercially available.

C.6 TRANSMISSION LINE INSULATORS

In order to prevent the flow of current from the transmission lines to the earth, the lines are secured to the supporting towers or poles by means of insulators.

The main requirements of a transmission line insulator are enumerated below:

(*i*) They must be mechanically strong.

(*ii*) They must have high dielectric strength.

(*iii*) They must provide high insulation resistance to leakage currents.

(*iv*) They must be free from internal impurities.

(*v*) They should not be porous.

(*vi*) They must be impervious to the entry of gases or liquids.

(*vii*) They should not be affected by changes in temperature or seasonal changes.

(*viii*) They must have a high puncture voltage to flashover voltage ratio.

(*ix*) They must have uniform creepage distance.

C.7 DEFINITIONS

1. *Flashover Voltage*: It is the minimum voltage at which the air around the insulator breaks down.
2. *Puncture Voltage*: It is the minimum voltage at which the insulator material breaks down. Puncture leads to permanent damage of the insulation material, which explains why the material must have a high puncture to flashover voltage ratio.
3. *Creepage Distance*: It is the shortest distance between the high voltage and ground electrodes measured over the surface of the insulator.

C.8 MATERIALS USED

1. **Porcelain:** It is manufactured from china clay, which is naturally available as aluminium silicate. It is mixed with plastic kaolin, feldspar and quartz and the mixture is heated in a kiln at a controlled temperature. The insulator so obtained must be hard, smooth and non-porous. The material is glazed so as to prevent the accumulation of dust and moisture on the surface of the insulator as their presence will reduce the flashover voltage.

If the insulating material is manufactured at lower temperatures, its mechanical properties improve, but the material remains porous and may deteriorate once pressed into service.

If manufactured at a higher temperature, the porosity of the material decreases, but the material becomes more brittle.

Thus a compromise is made between the mechanical strength and the porosity of the material and a suitable kiln temperature is decided upon.

A mechanically sound porcelain insulator has a dielectric strength of 60000 V per cm of its thickness.

2. **Glass:** The glass is made tough by annealing. The advantages and disadvantages of glass insulators compared to porcelain insulators are listed below:

Advantages:

(*i*) They have dielectric strengths of up to 140 kV per cm.

(*ii*) They have high resistivity when properly annealed.

(*iii*) They have low coefficient of thermal expansion.

(*iv*) Simpler designs are possible due to their inherently high dielectric strength.

(*v*) They have high compressive strength.

(*vi*) They are transparent, thus making detection of flaws easy.

(*vii*) They are homogeneous.

(*viii*) They are cheaper.

Disadvantages:

(*i*) Moisture easily condenses over the surface of the material, due to which dirt will deposit, thus allowing the flow of leakage currents.

(*ii*) For higher voltages, glass cannot be cast in irregular shapes as internal strains will be caused due to irregular cooling.

3. **Steatite:** Steatite is a magnesium silicate naturally available as magnesium oxide and silica. It has a very high tensile strength and can be used in situations where transmission lines take sharp turns.

C.9 TYPES OF LINE INSULATORS

- *Pin Type*:

They are screwed on and firmly attached to galvanized steel bolts. For higher voltages, more than one such insulator can be stacked together. Pin insulators can be used for voltages up to 50 kV, but they are never used for higher voltages, as use of suspension insulators will prove to be cheaper. The common practice is to use pin insulators for voltages up to 33 kV.

- *Suspension Insulators*:

With increase in line voltage, the pin insulators will become heavy and complicated. The use of pin insulators would also escalate the cost. Thus, for higher voltages, suspension insulators are used. A number of such insulators are connected in series by metallic links to form a chain and the line conductor is carried by the bottom most insulator.

The advantages of such a system are:

(*i*) Each insulator is designed for 11 kV. Thus, by connecting a number of such insulators, a string of insulators can be designed for any system voltage.

(*ii*) If one of the insulators in a string fails, it can be replaced at a very low cost.

(*iii*) Mechanical stress on the string is less.

(*iv*) Line conductors are less affected by lightning.

(*v*) If load to be supplied by the transmission line increases, the potential of the existing line can also be increased by simply increasing the number of insulators in the string.

There are three types of suspension insulator namely, Hewitt suspension type, Cement cap type and Core and Link type.

- *Strain Insulator*:

When there is a dead end of line or a sharp curve, the line has to withstand a great strain. In such a case strain insulators are used. Strain insulators are similar to suspension insulators except for the fact that while suspension insulators are placed in the vertical plane, strain insulators are generally placed in the horizontal plane.

- *Shackle Insulator*:

Shackle insulators are mostly used on low voltage distribution lines. They can be used in a horizontal or vertical position. The conductors in the groove are fixed with the help of soft binding wires.

- *Power Station Batteries*:

A battery is a device that converts chemical energy contained in its active materials directly into electrical energy by means of an oxidation-reduction electrochemical reaction. The total amount of energy that can be provided by such a cell is entirely dependent on the amount of active material.

A battery consists of one or more cells connected in series, parallel or series parallel depending on the desired output voltage and capacity.

The cell consists of three main component—anode or reducing material, cathode or oxidizing material and the electrolyte which provides the necessary internal ionic activity.

C.10 TYPES OF ELECTROCHEMICAL CELLS

- *Primary Cell*: It is a device by which the electrical energy is got from chemical energy through an irreversible reaction. Current is supplied till the electrolyte is exhausted or negative electrode is completely dissolved. The cell cannot be recharged and the electrolyte will have to be replaced after long use.
- *Secondary Cell*: Here, electrical energy is stored as chemical energy (charged cell) and when current is drawn from the cell, the stored chemical energy is converted into electrical energy. The chemical reaction is reversible in this case and such cells are also called storage cells or accumulators.
- *Cell Rating*: The rating assigned to a battery is the capacity in Ah (after correction to 27°) stated by the manufacturer that can be got when the battery is discharged at the 10 hour rate to an end voltage of 1.85 V per cell. This is known as 10 hour rating of the cell and is designated as C_{10}. Ratings of C_5 and C_{20} are also popularly used.

C.11 FACTORS RESPONSIBLE FOR BATTERY CAPACITY

(*i*) Final limiting voltage of battery.
(*ii*) Discharge rate of battery.
(*iii*) Numbers, design and dimension of plates.
(*iv*) Design of separator.
(*v*) Quantity, quality and density of electrolyte.
(*vi*) Temperature of cells.

REFERENCE

1. *"Electric Power Distribution"* by A.S.Pabla.

Appendix D NATURE OF LOADS

Power system analysis and modeling depends on the type of the load connected. For example, analysis of the primary section of a distribution feeder depends on a different type of load as compared to the steady state analysis of a transmission system.

D.1 TERMINOLOGY USED WHEN DESCRIBING LOADS

Following are some important terms that are used during describing loads:

1. **Demand**
   - Load averaged over a specific period of time
   - Load can be kW, kVAR, kVA, or A
   - Must include the time interval
   - Example: The 30-minute kW demand is 500 kW
2. **Maximum Demand**
   - Greatest of all demands that occur during a specific time
   - Must include demand interval, period, and units
   - Example: The 15-minute maximum kW demand for the week was 100 kW
3. **Average Demand**
   - The average of the demands over a specified period (day, week, month, etc.)
   - Must include demand interval, period, and units
   - Example: The 15-minute average kW demand for the month was 600 kW
4. **Diversified Demand**
   - Sum of demands imposed by a group of loads over a particular period

- Must include demand interval, period, and units
- Example: The 45-minute diversified kW demand in the period ending at 10:00 was 200 kW.

5. **Maximum Diversified Demand**
   - Maximum of the sum of the demands imposed by a group of loads over a particular period
   - Must include demand interval, period, and units
   - Example: The 25-minute maximum diversified kW demand for the week was 500 kW
6. **Demand Factor**
   - Ratio of maximum demand to connected load
7. **Utilization Factor**
   - Ratio of the maximum demand to rated capacity
8. **Load Factor**
   - Ratio of the average demand of any individual customer or group of customers over a period to the maximum demand over the same period.

D.2 LOAD MODELS

Loads are generally specified by the complex power consumed. The demand can be specified in kW and pf or kVA and pf or kW and kVAR. Loads are modeled as star connected or delta connected. They can be single phase, three phase or two phase with varying degrees of unbalance.

A. Star Connected Loads:

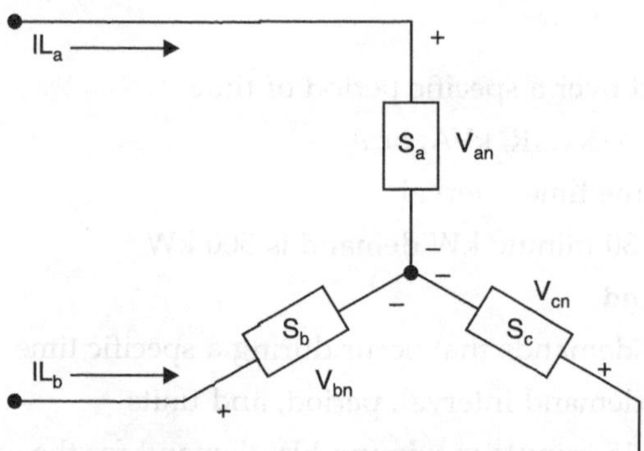

Phase (*a*): $|S_a|\underline{/\theta_a} = P_a + jQ_a$ and $|V_{an}|\underline{/\delta_a}$

Phase (*b*): $|S_b|\underline{/\theta_b} = P_b + jQ_b$ and $|V_{bn}|\underline{/\delta_b}$

Phase (*c*): $|S_c|\underline{/\theta_c} = P_c + jQ_c$ and $|V_{cn}|\underline{/\delta_c}$

(*a*) *Constant Real and Reactive Power Loads:*

$$IL_a = \left(\frac{S_a}{V_{an}}\right)^* = \frac{|S_a|}{|V_{an}|}\angle\delta_a - \theta_a = |IL_a|\angle\alpha_a$$

$$IL_b = \left(\frac{S_b}{V_{bn}}\right)^* = \frac{|S_b|}{|V_{bn}|}\angle\delta_b - \theta_b = |IL_b|\angle\alpha_b$$

$$IL_c = \left(\frac{S_c}{V_{cn}}\right)^* = \frac{|S_c|}{|V_{cn}|}\angle\delta_c - \theta_c = |IL_c|\angle\alpha_c$$

(*b*) *Constant Impedance Loads:*

$$Z_a = \frac{|V_{an}|^2}{S_a^*} = \frac{|V_{bn}|^2}{|S_a|}\angle\theta_a = |Z_a|\angle\theta_a$$

$$Z_b = \frac{|V_{bn}|^2}{S_b^*} = \frac{|V_{bn}|^2}{|S_b|}\angle\theta_b = |Z_b|\angle\theta_b$$

$$Z_c = \frac{|V_{cn}|^2}{S_c^*} = \frac{|V_{cn}|^2}{|S_c|}\angle\theta_c = |Z_c|\angle\theta_c$$

The load currents in such a model are given by :

$$IL_a = \frac{V_{an}}{Z_a} = \frac{|V_{an}|}{|Z_a|}\angle\delta_a - \theta_a = |IL_a|\angle\alpha_a$$

$$IL_b = \frac{V_{bn}}{Z_b} = \frac{|V_{bn}|}{|Z_b|}\angle\delta_b - \theta_b = |IL_b|\angle\alpha_b$$

$$IL_c = \frac{V_{cn}}{Z_c} = \frac{|V_{cn}|}{|Z_c|}\angle\delta_c - \theta_c = |IL_c|\angle\alpha_c$$

(*c*) *Constant Current Loads:*

$$IL_a = |IL_a|\angle\delta_a - \theta_a$$

$$IL_b = |IL_b|\angle\delta_b - \theta_b$$

$$IL_c = |IL_c|\angle\delta_c - \theta_c$$

δ-Line to Neutral voltage angles θ-Power factor angles

B. Delta Connected Loads:

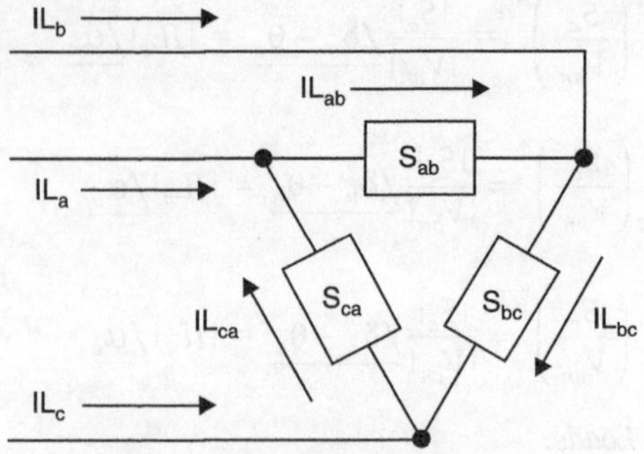

Phase (*ab*): $|S_{ab}|\underline{/\theta_{ab}} = P_{ab} + jQ_{ab}$ and $|V_{ab}|\underline{/\delta_{ab}}$

Phase (*bc*): $|S_{bc}|\underline{/\theta_{bc}} = P_{bc} + jQ_{bc}$ and $|V_{bc}|\underline{/\delta_{bc}}$

Phase (*ca*): $|S_{ca}|\underline{/\theta_{ca}} = P_{ca} + jQ_{ca}$ and $|V_{ca}|\underline{/\delta_{ca}}$

(a) Constant Real and Reactive Power Loads:

$$IL_{ab} = \left(\frac{S_{ab}}{V_{ab}}\right)^* = \frac{|S_{ab}|}{|V_{ab}|}\underline{/\delta_{ab} - \theta_{ab}} = |IL_{ab}|\underline{/\alpha_{ab}}$$

$$IL_{bc} = \left(\frac{S_{bc}}{V_{bc}}\right)^* = \frac{|S_{bc}|}{|V_{bc}|}\underline{/\delta_{bc} - \theta_{bc}} = |IL_{bc}|\underline{/\alpha_{bc}}$$

$$IL_{ca} = \left(\frac{S_{ca}}{V_{ca}}\right)^* = \frac{|S_{ca}|}{|V_{ca}|}\underline{/\delta_{ca} - \theta_{ca}} = |IL_{ca}|\underline{/\alpha_{ca}}$$

(b) Constant Impedance Loads:

$$Z_{ab} = \frac{|V_{ab}|^2}{S_{ab}^*} = \frac{|V_{ab}|^2}{|S_{ab}|}\underline{/\theta_{ab}} = |Z_{ab}|\underline{/\theta_{ab}}$$

$$Z_{bc} = \frac{|V_{bc}|^2}{S_{bc}^*} = \frac{|V_{bc}|^2}{|S_{bc}|}\underline{/\theta_{bc}} = |Z_{bc}|\underline{/\theta_{bc}}$$

$$Z_{ca} = \frac{|V_{ca}|^2}{S_{ca}^*} = \frac{|V_{ca}|^2}{|S_{ca}|}\underline{/\theta_{ca}} = |Z_{ca}|\underline{/\theta_{ca}}$$

The load currents are given as

$$IL_{ab} = \frac{V_{ab}}{Z_{ab}} = \frac{|V_{ab}|}{|Z_{ab}|}\underline{/\delta_{ab} - \theta_{ab}} = |IL_{ab}|\underline{/\alpha_{ab}}$$

$$IL_{bc} = \frac{V_{bc}}{Z_{bc}} = \frac{|V_{bc}|}{|Z_{bc}|}\underline{/\delta_{bc} - \theta_{bc}} = |IL_{bc}|\underline{/\alpha_{bc}}$$

$$IL_{ca} = \frac{V_{ca}}{Z_{ca}} = \frac{|V_{ca}|}{|Z_{ca}|}\underline{/\delta_{ca} - \theta_{ca}} = |IL_{ca}|\underline{/\alpha_{ca}}$$

(*c*) *Constant Current Loads:*

$$IL_{ab} = |IL_{ab}|\underline{/\delta_{ab} - \theta_{ab}}$$

$$IL_{bc} = |IL_{bc}|\underline{/\delta_{bc} - \theta_{bc}}$$

$$IL_{ca} = |IL_{ca}|\underline{/\delta_{ca} - \theta_{ca}}$$

(*d*) *Two Phase and Single Phase Loads:*

Single phase and two phase loads are modeled by setting the currents of the corresponding missing phases to zero. The above models equally apply to such loads.

D.3 APPROXIMATE LINE SEGMENT MODEL

This model is used when the only given line parameters are its positive and zero sequence impedances.

$$[Z_{\text{seq}}] = \begin{bmatrix} Z_0 & 0 & 0 \\ 0 & Z_+ & 0 \\ 0 & 0 & Z_+ \end{bmatrix}$$

Reverse impedance transformation is used to get :

$$[Z_{\text{approx}}] = [A_s] \,.\, [Z_{\text{seq}}] \,.\, [A_s]^{-1}$$

$$[Z_{\text{approx}}] = \frac{1}{3} \cdot \begin{bmatrix} (2 \,.\, Z_+ + Z_0) & (Z_0 - Z_+) & (Z_0 - Z_+) \\ (Z_0 - Z_+) & (2 \,.\, Z_+ + Z_0) & (Z_0 - Z_+) \\ (Z_0 - Z_+) & (Z_0 - Z_+) & (2 \,.\, Z_+ + Z_0) \end{bmatrix}$$

where $[A] = \begin{bmatrix} 1 & 0 & 0 \\ 0 & 1 & 0 \\ 0 & 0 & 1 \end{bmatrix}$

REFERENCE

1. S. Z. Zhu Z. Y. Dong K. P. Won Z. H. Wang, "*Power System Dynamic Load Identification and Stability*" 0-7803-6338-8/00, 2000 IEEE page no. 13–19.

INDEX

A

B

C

D

E

F

G